PASS YOUR AMATEUR RADIO EXTRA CLASS TEST – THE EASY WAY
2020-2024 Edition

By: Craig E. "Buck," K4IA

ABOUT THE AUTHOR: "Buck," as he is known on the air, was first licensed in the mid-sixties as a young teenager. Today, he holds an Amateur Extra Class Radio License. Buck is an active instructor and a Volunteer Examiner. The Rappahannock Valley Amateur Radio Club named him Elmer (Trainer) of the Year three times. Buck is on the DXCC Honor Roll and has an 8 Band DXCC endorsement.

Email: K4IA@EasyWayHamBooks.com

Published by EasyWayHamBooks
130 Caroline St. Fredericksburg, Virginia 22401

Easy Way Ham Books by Craig Buck are available at Ham Radio Outlet stores and on Amazon:
Pass Your Amateur Radio Technician Class Test
Pass Your Amateur Radio General Class Test
How to Chase, Work & Confirm DX
How to Get on HF
Pass Your GROL Test
Prepper Communications

Copyright ©2020, Craig E. Buck   All Rights Reserved. No part of this material may be reproduced, transmitted or stored in any manner, in any form or by any means without the express written permission of the author. v3.4 Note: This version includes the VEC issued errata March 19, 2020

ISBN 9798614382407
Library of Congress Control Number: 2020902271

PASS YOUR AMATEUR RADIO EXTRA CLASS TEST – THE EASY WAY

TABLE OF CONTENTS

TABLE OF CONTENTS i
INTRODUCTION 1
THE TEST 2
HOW YOU SHOULD STUDY 6
E1 – COMMISSION'S RULES 7
   E1A Operating Standards 7
   E1B Station Restrictions 9
   E1C Control 11
   E1D Satellites 12
   E1E Examiners 14
   E1F Miscellaneous Rules 16
E2 – OPERATING PROCEDURES 18
   E2A Amateur Radio in Space 18
   E2B Television 20
   E2C Operating Methods; Contest and DX 21
   E2D Operating Methods: VHF/UHF 23
   E2E Operating Methods: HF digital 24
E3 – RADIO WAVE PROPAGATION 27
   E3A Electromagnetic Waves 27
   E3B Transequatorial Propagation 28
   E3C Radio Propagation 30
E4 – AMATEUR PRACTICES 32
   E4A Test Equipment 32
   E4B Measurements 33
   E4C Receiver Performance 35
   E4D Receiver Performance 38
   E4E Noise Suppression and Grounding 39
E5 – ELECTRICAL PRINCIPLES 42
   E5A Resonance and Q 42
   E5B Time and Phase 44
   E5C Coordinate Systems 46

E5D AC and RF Energy 49
E6 – CIRCUIT COMPONENTS 52
E6A Semiconductors 52
E6B Diodes 54
E6C Digital ICs 55
E6D Torroidal and Solenoidal Inductors; 57
E6E Analog ICs; MMICs; IC Packaging 59
E6F Optical Components 61
E7 – PRACTICAL CIRCUITS 63
E7A Digital Circuits 63
E7B Amplifiers 64
E7C Filters and Matching Networks 67
E7D Power Supplies 69
E7E Modulating and Demodulating 71
E7F DSP and Software-Defined Radio 73
E7G– Active Filters and Op-amp Circuits 75
E7H Oscillators and Signal Sources 77
E8 – SIGNALS AND EMISSIONS 80
E8A Waveforms..80
E8B Modulation and Demodulation 81
E8C Digital Signals 83
E8D Keying Defects and Overmodulation 84
E9 – ANTENNAS AND TRANSMISSION LINES 87
E9A Basic antenna parameters 87
E9B Antenna Patterns 89
E9C Wire and Phased Array Antennas.................92
E9D Directional and Short Antennas...................94
E9E Matching 95
E9F Transmission Lines 97
E9G Smith Chart 100
E9H Receiving Antennas 102
E0 – SAFETY 104
E0A Safety 104
AMATEUR RADIO EXTRA CLASS QUICK SUMMARY 106
E1 — COMMISSION RULES 106
E1A Operating Standards 106
E1B Station Restrictions 108
E1C Control 110
E1D Satellites 111

Page ii Extra Class – The Easy Way

E1E Examiners 113
E1F Miscellaneous Rules 115
E2 - OPERATING PROCEDURES.......................117
E2A Amateur Radio in Space 117
E2B Television..118
E2C Operating Methods: Contest and DX 120
E2D Operating methods: VHF and UHF 121
E2E Operating Methods: Digital 123
E3 - RADIO WAVE PROPAGATION....................125
E3A Electromagnetic Waves 125
E3B Transequatorial Propagation 126
E3C Radio Propagation 127
E4 - AMATEUR PRACTICES............................130
E4A Test Equipment..130
E4B Measurements 131
E4C Receiver Performance 133
E4D Receiver Performance 135
E4E Noise Suppression and Grounding..............136
E5 – ELECTRICAL PRINCIPLES........................139
E5A Resonance and Q 139
E5B Time and Phase 141
E5C Coordinate Systems..................................142
E5D AC and RF Energy 143
E6 - CIRCUIT COMPONENTS 146
E6A Semiconductors 146
E6B Diodes 147
E6C Digital ICs 148
E6D Toroidal and Solenoidal Inductors..............150
E6E Analog ICs: MMICs, IC packaging 151
E6F Optical Components..................................153
E7 – PRACTICAL CIRCUITS............................155
E7A Digital Circuits 155
E7B Amplifiers...156
E7C Filters and Matching Networks 158
E7D Power Supplies 159
E7E Modulation and Demodulation 161
E7F DSP and Software Defined Radio 163
E7G Active filters and Op-amp Circuits 164
E7H Oscillators and Signal Sources 166

Extra Class – The Easy Way   Page iii

E8 - SIGNALS AND EMISSIONS 168
- E8A Waveforms 168
- E8B Modulation and Demodulation 169
- E8C Digital Signals 170
- E8D Keying defects and Overmodulation 172

E9 - ANTENNAS AND TRANSMISSION LINES ..174
- E9A Basic Antenna Parameters 174
- E9B Antenna Patterns 175
- E9C Wire and Phased Array Antennas 177
- E9D Directional and Short Antennas 178
- E9E Matching 180
- E9F Transmission Lines 181
- E9G Smith Chart 183
- E9H Receiving Antennas 185

E0 – SAFETY 187
- E0A Safety 187

BONUS MATERIAL – LEARNING CW 194

INDEX 197

Note: This version includes the errata issued by the Volunteer Examiner Committee on March 19, 2020

INTRODUCTION

There are many books and methods to help you study for the amateur radio exams. Most focus on taking you through all the questions and possible answers on the multiple-choice test. The problem with that approach is that you must read three wrong answers for every one right answer. That's 1,866 wrong answers and 622 correct answers. No wonder people get overwhelmed.

This book is different. There are no confusing wrong answers. I'll answer every question on the amateur radio Extra Class exam in a narrative format. **I've put the test questions and answers in bold print to help you focus.** *Hints to help you decipher the questions and answers are in italic print. Cheats give you a shortcut memory jogger to the answer.*

The questions are roughly in the order they appear in the question pool. I have rearranged a few to keep topics together. Please excuse any tortured grammar or bad punctuation in the bold print questions and answers. I often copy them word-for-word from the pool.

The second part of this book is a Quick Summary; no narrative, only the question, and the right answer.

You will pass your Extra Class test, but it won't tell you much about how to operate, chose equipment and put up antennas. Be sure to check out my other books: "**How to Get on HF – The Easy Way**" for detailed instructions to get on the air and "**How to Chase, Work & Confirm DX – The Easy Way**."

THE TEST

This book is for tests after June 30, 2020 and before July 1, 2024. The test is 50 questions out of a possible pool of 622. The pool is large, but only about one of a dozen questions in the pool will show up on your exam - one from each test-subject grouping. You won't get 50 questions on one topic. Some questions ask for the same information in a slightly different format, so there aren't 622 unique questions.

The Extra test is harder than the General, but you can do it. Don't be intimidated. This book will teach you to recognize the correct answer – even if you don't understand the question. Go through this material with the intent to identify the correct answer.

There is no Morse Code required for any class of amateur license. Morse Code is still very much alive and well on the amateur bands, and there are many reasons you might want to learn it in the future.[1]

The good news is that we know the exact pool of questions and answers. If "200" is the answer in the question pool, "200" will also be the answer to that question on your exam. Answer "A" may not be the same as answer "A" on your test, so don't try to memorize the answer's letter.

You only get one question from each group. Do your best, but if there is a question or concept you can't understand, don't give up. Chances are it won't be on your test. Ninety-two percent of the questions in the pool will not be on your test. You can also visit my Facebook group "Ham Radio Exams" and ask for help.

---

[1] Check out my book "How to Chase, Work and Confirm DX – The Easy Way." I discuss the advantages of CW over voice modes in detail.

THE TEST

You must answer thirty-seven out of the fifty questions on your exam to pass. **The minimum passing score is 74%.** (There are your first question and answer). There are 622 questions in the pool. Have the correct answer to 6% of the pool, and you pass. Feel better?

Most important: when the day comes, take the test, even if you feel you are not ready. You will be better prepared than you think.

Do you know what they call the person who graduates last in his class from medical school? "Doctor" – the same as the one who graduated first in the class. My point is: pass, and no one will know the difference.

The questions are multiple-choice, so you don't have to know the answer; you have to <u>recognize</u> it. That is a tremendous advantage for the test-taker. However, it's challenging to study the multiple-choice format because you can get bogged down and confused by seeing three wrong answers for every correct one.

The best way to study for a multiple-choice exam is to concentrate on the correct answers. That is the focus of this book. When you take the test, the correct response should jump out at you. The wrong answers will seem strange and unfamiliar, as if in a foreign language.

Don't over-think or over-analyze. You should recognize the right answer without thinking. If you don't remember the answer, try to eliminate the obviously wrong answers and then guess. There is no penalty for guessing wrong.

Here's the ultimate secret cheat: many times, the answer is a matter of common sense and logic. Often the question gives away the answer. Many of the hints and cheats in this book exploit this.

THE TEST

The test-day routine is the same as it was for your General test. You should bring a picture ID and your FRN Number. The FRN is used only for your communications with the FCC. Search "FRN FCC" to get the link to the FCC website where you can register. Then, bring your FRN number to the test session. If you are already licensed, the FCC has assigned you an FRN. Look it up on FCC.Gov and search in the ham database for your last name, first name, or call sign.

Bring a pen and pencil and the exam fee (usually around $15). The FCC has proposed a $35 licensing fee, but as of July 2021, this is not being collected. Check ahead to get the exact amount and if the VE team prefers cash or a check.

If you bring a calculator, you must clear the memories. Turn off your cellphone and don't look at it during the exam. You won't be able to use the calculator on your cellphone because the VEs can't be sure you aren't looking for answers on the Internet.

You will fill out a simple application (FCC Form 605). Get one from the FCC website and fill it out beforehand to save time on test day. It will require an email address that becomes your official contact medium for the FCC.

The Volunteer Examiner team will give you a test booklet. Ask them for the easy one – that's good for a cheap laugh. They have no idea which questions are in which booklet. You also get a multiple-choice, fill-in-the-circle answer sheet. You can take notes and calculate on the back of the answer sheet – not in the booklet. The VEs re-use the booklets.

You can do a "memory dump" on the back of the answer sheet before you get started. Refer to it later if you get confused.

THE TEST

The VE team will grade your exam while you wait and give you a pass/fail result. Don't ask them to go over the questions or tell you what you missed. They don't know and don't have time to look. There is a bit more paperwork when you pass, and you walk out with a CSCE – a Certificate of Successful Completion of Examination.

> You may not need to take the Extra Class test. If you are beyond the two-year grace period for a test-free renewal but can demonstrate you once held an Amateur Extra class license, which the FCC did not revoke, you can get credit. You must only pass the current Technician Class test. Check with your local VE team for details.

If you are already licensed, once you have the CSCE for passing Extra, you can operate as an Extra even though you don't yet show up in the FCC database.

When you use Extra Class privileges, use the special identifier "AE" after your call sign while waiting for the FCC to post your upgrade on its website. That would be W3ABC/AE. W3ABC stroke/slash/slant AE. *Hint: Awaiting Extra.* After you appear in the database, drop the AE.

The FCC no longer routinely mails out paper licenses or notices. You can print an "Official Copy" of your license or request one from the FCC website. The email address you gave on Form 605 is your official point of contact. Keep it current as you are presumed to receive anything sent to that address, and failure to respond could result in license cancellation.

HOW YOU SHOULD STUDY

You ace a multiple-choice exam by learning to recognize the correct answers and eliminating wrong answers. You could study by pouring over the multiple-choice questions. That has been the traditional way of most classes and license manuals. The problem with that method is you have to read through three wrong answers to every question. That is both frustrating and confusing. Why study the WRONG answers?

The approach here is that you never see the wrong answers, so the correct answer should pop out when you see it on the test. You don't need to memorize the whole answer – just enough of it to recognize. Many of the questions and answers are long and involved. They are full of information you don't need to get the correct answer. Sometimes, all you need to do is tie two words together. Look for those.

I don't recommend you take practice exams until the week before the test after you feel you have mastered this material. I suggest you wait to take the practice exams because I don't want you to get confused by seeing all those wrong answers. Take the practice test to build your confidence but study this book. Recognize the correct answers first.

If you are stuck on a question or concept, visit my Facebook group "Ham Radio Exams." You will find plenty of folks happy to assist and encourage.

E1 – COMMISSION'S RULES

E1A Operating Standards[2]

Your signal is wider than you think, so you have to be careful not to get too close to the band edges. That is the lesson of a series of questions.

If you are operating on LSB, your signal is all below the displayed frequency. **If you are operating LSB AFSK in the 17-meter band data segment of 18.068 to 18.110 MHz, it would be illegal to transmit on 18.068 MHz.** Your LSB signal would extend below the bottom frequency. *Hint: The question gives away the answer.*

An SSB phone signal is about 3 kHz wide. **The displayed carrier frequency should be 3 kHz above the lower band edge if using LSB.** Your LSB signal will take up 3 kHz below the carrier.

The maximum legal carrier frequency on the 20-meter band using USB digital signals having a 1 kHz bandwidth would be 14.149 MHz. *Hint: Too complicated. Remember, the upper edge of the digital sub-band is 14.150. If your USB signal is 1 kHz wide, you need to be below 14.149.*

If you hear a station calling CQ on 3.601 MHz LSB, it is NOT legal to return the call because your sideband would extend beyond the band edge. The band edge for voice is 3.6 MHz. Your LSB signal would extend below that.

---

[2] (E1A) refers to the question pool subelement. You get one question from each subelement. 50 Subelements, 50 questions.

COMMISSION'S RULES

We share the 60-meter bands with other services, so it has different rules. **The maximum power permitted on 60 meters is 100 watts effective radiated power as compared to a half-wave dipole.** Hint: Remember 100 watts and a dipole on 60 meters.

On 60 meters, set your CW carrier frequency at the center frequency of the channel. Sixty-meters is the only band restricting amateurs to channels.

The maximum power permitted on the 2200-meter band is 1 watt EIRP (Equivalent isotropic radiated power).

The maximum power permitted on the 630-meter band is 5 watts EIRP.

Phone emissions are permitted in the entire 630-meter band.

Message forwarding stations (like Packet) are usually automated. **If a station forwards a message that violates FCC rules, the control operator of the originating station is accountable for the rules violation.**

If your digital message forwarding station inadvertently forwards a communication that violated FCC rules, you should discontinue forwarding the communication as soon as possible.

Slow scan TV is restricted to phone band segments. Slow scan is about 3 kHz wide, the same as an SSB phone signal.

You can operate from a ship or airplane, but you must obtain permission from the master of the

COMMISSION'S RULES

ship or pilot in command of the aircraft. You need permission from the boss.

On a US registered vessel in international waters, any FCC-issued amateur license will do. No special endorsements required.

Any person holding an FCC-issued amateur radio license or who is authorized for alien reciprocal operation can be in physical control of an amateur station on a vessel or craft registered in the United States. *Hint: Look for "FCC-issued amateur radio license" in both answers."*

E1B Station Restrictions

A spurious emission is an emission outside its necessary bandwidth that can be reduced or eliminated without affecting the information transmitted. *Hint: Spurious is outside its necessary bandwidth.*

An acceptable bandwidth for Digital Radio Mondiale (DRM) based voice is 3 kHz. Digital Radio Mondiale is digital audio. It should be about the same width as SSB.

You must protect an FCC monitoring facility from harmful interference if you are within 1 mile.

Before placing an amateur station within an officially designated wilderness area or wildlife preserve or an area listed in the National Register of Historic Places, you must submit an Environmental Assessment to the FCC.

The National Radio Quiet Zone is an area surrounding the National Radio Astronomy Observatory. *Cheat: "National is in both the question and answer.*

COMMISSION'S RULES

Before installing an antenna at a site near a public use airport, you may have to notify the Federal Aviation Administration and register it with the FCC.

PRB-1 applies to state and local zoning. State and local governments must "reasonably accommodate" amateur radio. No such restrictions apply to homeowner associations. They can be unreasonable.

If an amateur station's signal causes interference to domestic broadcast reception, the FCC may place limitations to avoid transmitting during certain hours on frequencies that cause the interference. The FCC can impose "quiet times."

RACES is the Radio Amateur Civil Emergency Service and is part of a protocol established by FEMA and the FCC. **Any FCC-licensed amateur station certified by the responsible civil defense organization for the area served may be operated under RACES rules.** Hint: RA<u>C</u>ES <u>c</u>ertifies.

The frequencies authorized for amateur stations under RACES rules are all amateur service frequencies authorized to the control operator. Nothing special.

If a repeater interferes with a radio location system, the control operator must cease operation or make changes to mitigate the interference. It makes sense that you can't interfere with a radio location system.

COMMISSION'S RULES

E1C Control

The maximum bandwidth for data emissions on 60 meters is 2.8 kHz. About the same as SSB voice.

Communications incidental to the purpose of the amateur service and remarks of a personal nature are the types of communication that may be transmitted to amateur stations in foreign countries. Same as domestic.

An IARP is an International Amateur Radio Permit that allows US amateurs to operate in certain countries of the Americas. It is a multi-country license.

The control operator responsibilities under automatic control are different from under local control because, under automatic control, the control operator is not required to be present at the control point.

An automatically controlled station may never originate third party communications.

The maximum allowed duration of a remotely controlled station's transmissions if its control link malfunctions is 3 minutes. It must time-out after 3 minutes.

CEPT is an agreement for US amateurs to operate in European countries and vice-versa.

To operate within the CEPT rules, you must bring a copy of FCC Public Notice DA 16-1048. It lists the countries and rules in several languages. *Cheat: Just remember "FCC Public Notice."*

A signal's bandwidth is determined 26 dB below its mean power level.

Extra Class – The Easy Way   Page 11

COMMISSION'S RULES

The highest modulation index permitted for angle modulation below 29 MHz is 1. Angle modulation is FM. The FM Modulation index is equal to the ratio of the frequency deviation to the modulating frequency. A ratio of 1 is narrow.

The permitted mean power of any spurious emission is 43 dB below the fundamental emission. *Hint: The question is too complicated. Remember, spurious emissions must be 43 dB below.*

Phone emissions are permitted in the entire 630-meter band.

Power-line communication (PLC) carries data on power lines used for electric power distribution to consumers. Operation on 630 or 2200 meters could interfere. **Before operating in the 630 or 2200-meter band, you must notify the Utilities Technologies Council of your call sign and coordinates of your station.** *Hint: You never heard of the Utilities Technologies Council before, it is the correct answer.*

You must wait 30 days after notifying UTC before you can transmit, provided you have not been told your station is within 1 km of a PLC system using those frequencies. *Hint: Too complicated. Look for the answer with "wait 30 days."*

E1D Satellites

Telemetry is one-way transmission of measurements. *Hint: Telemetry is metering.*

Telecommand signals from a space telecommand station may transmit special codes to obscure the meaning of a message. The beeps and pulses don't mean anything to a listener.

COMMISSION'S RULES

A space telecommand station is a station that transmits communications to initiate, modify, or terminate functions of a space station. *Hint: It commands the space station to do something.*

A balloon-borne telemetry station must identify with a call sign. Like everyone else.

A station being operated by telecommand on or within 50 km of the Earth's surface must post at the station location:
- A photocopy of the station license
- A label with the name, address and telephone number of the station licensee
- A label with the name address and telephone number of the control operator.
- **All these choices are correct.**

All this information would help find the owner.

The maximum power permitted when operating a model craft by telecommand is 1 watt.

The HF bands authorized for space stations are 40 m, 20 m, 17 m, 15 m, 12 m, and 10 m bands. *Hint: All of 40-10 except 30 m.*

On VHF, only 2 meters is available for space stations.

On UHF, 70 cm and 13 cm are available for space stations.

Any amateur station designated by the space station licensee is eligible to be a telecommand station. *Hint: You need permission from the boss, the space station licensee.*

Stations eligible to operate as Earth stations are any amateur station, subject to the privileges of the control operator.

COMMISSION'S RULES

The stations that may transmit one-way communications are a space station, beacon station or telecommand station. *Hint: Look for the answer with telecommand, one-way control communications.*

E1E Examiners

Count yourself fortunate that you do not have to travel to an FCC Field Office to take your test. The FCC turned testing responsibilities over to Volunteer Examiner Coordinators (VECs) in 1984. There are about 14 Volunteer Examiner Coordinator organizations, and they jointly write the question pools and administer the system. On test day, the folks you see are Volunteer Examiners (VEs), accredited by the Coordinators (VECs).

VEs may be reimbursed for expenses preparing, processing, administering and coordinating exams. Not for teaching or training.

The questions for all written U.S. amateur license examinations are listed in pools of questions maintained by the VECs. The Examiner Coordinator organizations maintain the pools, not the individual examiners.

The Volunteer Examiner Coordinator is an organization that has entered into an agreement with the FCC to coordinate, prepare and administer amateur radio license examinations. *Hint: A coordinator coordinates.*

The Volunteer Examiner accreditation process is the procedure by which a VEC confirms that the VE applicant meets FCC requirements. *Hint: A VE is accredited by a VEC, not the FCC.*

The minimum passing score on all exams is 74%.

COMMISSION'S RULES

Each administering VE is responsible for the proper conduct and necessary supervision during an amateur radio license examination session. Everyone is responsible for maintaining the integrity of the system. We don't want the FCC to take examinations back in-house.

If a candidate fails to comply with the examiner's instructions, the examiner should immediately terminate the exam.

A VE may not administer an examination to someone who is a relative of the VE as listed in the FCC rules. *Hint: Friends and employees are OK.*

A VE who fraudulently administers or certifies an examination can lose his amateur station license and amateur operator license. No fine or jail time – loss of license is worse!

After administering a successful examination, the VEs must submit the application to the coordinating VEC according to the VEC's instructions. The test is graded on-site by the VEs who send the results to their VEC. The VEC loads the data in the FCC database.

When an examinee scores a passing grade, three VEs must certify that the examinee is qualified for the license grant and that they have complied with the VE requirements. *Hint: Look for "three VEs," and that is the answer.*

If the examinee does not pass the exam, the VE team will return the application form to the examinee. You get the application back, not the fee.

Extra Class – The Easy Way   Page 15

COMMISSION'S RULES

E1F Miscellaneous Rules

Spread-spectrum transmissions change frequency in a deliberate pattern to decrease interference and assure privacy. The result is a wide signal, so it is limited to the higher and wider bands. **Spread-spectrum transmissions are permitted only on amateur frequencies above 222 MHz.**

A Canadian license holder is allowed the same privileges in the US, not to exceed the US Amateur Extra privileges.

Amplifiers must be FCC certified to protect against use by CBers and to assure spectrum purity. **A dealer may sell an external RF power amplifier capable of operation below 144 MHz if it has not been granted FCC certification if it was purchased in used condition from an amateur operator and is sold to another amateur operator for use at that operator's station.** *Hint: Too complicated! Just remember, amateur to amateur is OK.*

Line A is a line roughly parallel to and south of the US-Canadian border.

Amateur stations in the US may not transmit on 420 MHz – 430 MHz if they are north of Line A.

A Special Temporary Authority may be issued by the FCC to provide for experimental amateur communications. These usually allow for operation outside normal amateur frequencies.

An amateur station may send a message to a business only when neither the amateur nor his employer has any pecuniary interest in the communications.

COMMISSION'S RULES

Communications transmitted for hire or material compensation, except as otherwise provided in the rules, are prohibited. *Hint: "For hire" is prohibited. You can't be paid for being a Ham.*

Amateur radio isn't for business. Here's an example: I'm on my way over to my friend Bob's house to help him with an antenna. I call him on the radio and ask if he would like me to bring over pizza. He says, sure – whatever kind you like. Another ham, who happens to own a pizza parlor, hears me and calls. "Hey Buck, what kind of pizza do you want. I'll have it ready for you." That's a no-no: the same pizza but a different result.

If you are going to use spread spectrum, the following conditions apply:
- **Must not cause harmful interference**
- **Must be in an area regulated by the FCC or in a country that permits SS emissions**
- **Must not be used to obscure the meaning**
- **All of the choices are correct**

An auxiliary station is one that controls another station over a radio link. **The control operator of an auxiliary station may be a Technician, General, Advanced, or Amateur Extra operator.** Advanced is an old license class. *Look for the only answer that includes "Technician."*

To qualify for a grant of FCC certification, an amplifier must satisfy the FCC's spurious emissions standards when operated at 1500 watts or its full output power
Hint: To get certification, satisfy the FCC. Don't bother memorizing the power.

E2 – OPERATING PROCEDURES

E2A *Amateur Radio in Space*

The ascending pass for an amateur satellite is from south to north. *Cheat: "Ascending." Think of it as going up from the South Pole to the North.*

A transponder receives a signal and emits a signal in response. A satellite might receive a signal on one band (uplink) and retransmit on another (downlink).

When a satellite is using an inverting linear transponder:
- **Doppler shift is reduced because the uplink and downlink shifts are in opposite directions**
- **Signal position in the band is reversed**
- **Upper sideband on the uplink becomes lower sideband on the downlink, and vice versa.**
- **All these choices are correct.**

Hint: Everything in the answers is inverted.

The signal is inverted by an inverting linear transponder because the signal is passed through a mixer, and the difference rather than the sum is transmitted. *Hint: Inverted.*

A satellite's mode is the uplink and downlink frequency bands.
The letters in a satellite's mode designator specify the uplink and downlink frequencies.
If a satellite is operating in U/V mode, it means the uplink is UHF, and the downlink is VHF.

Keplerian elements are parameters that define the orbit of a satellite. *Hint: Kepler developed formulas for describing an orbiting body.*

OPERATING PROCEDURES

A linear transponder can relay:
- **FM and CW**
- **SSB and SSTV**
- **PSK and Packet**
- **All these choices are correct**

Hint: The transponder is linear, so it can relay any type of signal.

The effective radiated power to a satellite that uses a linear transponder should be limited to avoid reducing the downlink power to all other users. A satellite receiving a very strong signal assumes an excellent connection and will reduce its transmit power to conserve energy.

The terms "L band" and "S band" refer to the 23-centimeter and 13-centimeter bands. *Cheat: Remember 23 or 13, and you have the answer.*

A satellite that stays in one position in the sky is geostationary. *Hint: If it stays in one place, it is stationary.*

To minimize the effect of spin modulation and Faraday rotation, use a circularly polarized antenna. *Cheat: Spin in circles.*

The purpose of digital store and forward functions is to store digital messages in the satellite for later download by other stations. *Hint: Digital store and forward stores digital messages and forwards them later.*

Low Earth orbiting digital satellites relay messages around the world by store-and-forward. *Hint: "Around the world" requires storing the message until they get over the target area.*

OPERATING PROCEDURES

E2B Television

Fast-scan (NTSC) television transmits 30 frames per second.
Fast-scan (NTSC) television transmits 525 lines per frame. Fast-scan is 30 frames of 525 lines each.

An interlaced scanning pattern is generated in a fast-scan (NTSC) television by scanning odd numbered lines in one field and even numbered lines in the next. Hint: Odd and even are interlaced.

Color information in analog SSTV is sent by color lines sent sequentially — one color after the other.

Vestigial sideband modulation is amplitude modulation in which one complete sideband and a portion of the other are transmitted. Hint: Vestigial means a small remnant – a portion of the other sideband.

The advantage of vestigial sideband for standard fast-scan television is that it reduces bandwidth while allowing for simple video detector circuitry. Because only part of the other sideband is transmitted, it reduces bandwidth. Hint: "Reduces bandwidth."

The signal component that carries color information is called chroma. Hint: Chromatic means of, relating to, or produced by color.

The brightness of the picture is encoded by a tone frequency.

The technique that allows commercial analog TV receivers to be used for fast-scan TV operation in the 70 cm band is transmitting on channels shared with cable TV. Hint: Too much information!

OPERATING PROCEDURES

Commercial TV receivers can pick up cable TV.

A receiver with SSB capability and a suitable computer is needed to decode SSTV using Digital Radio Mondiale (DRM), and no other hardware is needed. DRM is decoded by a computer.

The Vertical Interval Signaling (VIS) code is sent as part of an SSTV transmission to identify the SSTV mode being used. *Hint: It is signaling the mode.*

SSTV receiving software is signaled to begin a new line by specific tone frequencies.
Hint: Look for "tone frequencies" in answers.

E2C Operating Methods; Contest and DX

When a U.S. licensed operator is operating a remote control transmitter in the U.S., no additional indicator is required (when identifying).

Self-spotting is the often prohibited practice of posting your own callsign and frequency on the spotting network. You can't spot yourself when contesting.

Amateur contesting is generally excluded on 30 meters. Contesting is not allowed on any of the WARC bands (30, 17, 12 meters).

A mesh network is interconnected wireless points using a wireless network adapter. **The frequencies used for mesh networks are shared with various unlicensed wireless data services.** *Hint: Wireless network adapters are unlicensed.*

Extra Class – The Easy Way

OPERATING PROCEDURES

The equipment commonly used to implement a ham radio mesh network is a standard wireless router running custom software.

The technique individual nodes use to form a mesh network is discovery and link establishment protocols. *Hint: A mesh network discovers other nodes and establishes links*

A DX QSL manager handles the receiving and sending of confirmation cards for a DX station.[3]

The U.S. QSL bureau system is used for contacts between a U.S. station and a non-U.S. station. The U.S. QSL bureau system does not handle domestic confirmations.

During a VHF/UHF contest, you would expect to find the highest level of activity in the weak signal segment of the band with most of the activity near the calling frequency. *Hint: Most people will listen for weak signals near the weak-signal calling frequency.*

Cabrillo format is the standard for submission of electronic logs. Your computer logging program will convert your file to Cabrillo format for uploading to the contest sponsor.

A DX station might state they are listening on a different frequency:
- **Because they are transmitting on a frequency prohibited to some of the responding stations.**

---

[3] Lots more details in my book "How to Chase, Work & Confirm DX – The Easy Way"

OPERATING PROCEDURES

- **To separate the calling stations from the DX station.**
- **To improve the operating efficiency by reducing interference.**
- **All these choices are correct.**

Many DX stations operate "split," listening on a different frequency from their transmissions. Get my book for details, "How to Chase, Work and Confirm DX. – The Easy Way."

When attempting to contact a DX station during a contest or pileup, you would generally identify by sending your full call once or twice. No partial calls, no grid squares, no repetitive identifying.

E2D Operating Methods: VHF/UHF

The digital mode designed for meteor scatter is MSK144. *Hint: Meteor SKatter*

A good technique for making meteor scatter contacts is:
- **15 second timed transmission sequences with stations alternating based on location.**
- **Use of high-speed CW or digital modes.**
- **Short transmission with rapidly repeated call signs and signal reports.**
- **All the choices are correct.**

Hint: Lots of good techniques, so all are correct.

The method of establishing EME contacts is time synchronous transmissions alternatively from each station. *Hint: To establish contact you need to know when to listen (time synchronous).*

The digital mode especially useful for EME communications is JT65. EME is Earth-Moon-Earth, and since neither moves very fast, you can use a slow mode, like JT65, designed for extremely weak signals.

OPERATING PROCEDURES

The type of modulation used for JT65 contacts is multi-tone AFSK. AFSK is Audio Frequency Shift Keying. *Hint: Remember JT65 uses audio tones.*

An advantage of JT65 mode is it can decode signals with a very low signal to noise ratio. *Hint: JT65 decodes signals below the noise.*

The technology used to track, in real time, balloons carrying amateur radio transmitters is APRS. APRS is Automatic Packet Reporting System. A GPS interfaces with your radio to send and receive position reports.

The digital protocol used by APRS is AX.25.

The type of packet frame used to transmit APRS beacon data is Unnumbered Information. *Hint: The data is information.*

An APRS station can help support public service communications because a GPS unit can automatically transmit information to show a mobile station's position during the event.

The data used by an APRS network to communicate your position is latitude and longitude. *Hint: Latitude and longitude show your position.*

E2E Operating Methods: HF digital

A common type of modulation for data emissions below 30 MHz is FSK. Frequency Shift Keying. FSK is used for RTTY (teletype). FSK is not the only type used for data emissions, but it is the only answer that is correct. *Hint: Recognize that FSK is a digital mode.*

OPERATING PROCEDURES

The letters FEC mean Forward Error Correction. Redundant data is sent, and if it doesn't match, a re-send request follows.

The timing of FT4 contacts is organized by alternating transmission at 7.5 second intervals. FT4 is a data mode that transmits for 7.5 seconds, then listens for 7.5 seconds.

If one of the ellipses in an FSK crossed-ellipse display suddenly disappears, selective fading has occurred. *Hint: The ellipse faded away.*

The digital mode that does NOT support keyboard-to-keyboard operation is PACTOR. PACTOR is one-way. A full message is transmitted without interruption. It is not a chat room.

The most common data rate for HF packet is 300 baud.
The digital mode with the fastest data throughput under clear communication conditions is 300 baud packet.
Hint: Look for "300 baud" in an answer.

If you are trying to initiate contact with a digital station on a clear frequency but are unsuccessful, the reason might be:
- **Your transmit frequency is incorrect.**
- **The protocol version you are using is not supported by the digital station.**
- **Another station you are unable to hear is using the frequency.**
- **All these choices are correct.**

Hint: Lots of reasons it might not work. If you recognize two, "all of the above" is the answer.

The HF digital mode used to transfer binary files is PACTOR. PACTOR sends and receives digital information using radio.

OPERATING PROCEDURES

PSK31 uses variable-length coding for bandwidth efficiency. The letters have different lengths. Capital letters take twice as long to send, so don't use all caps.

PSK31 has the narrowest bandwidth (31Hz)

The difference between direct FSK and audio FSK is direct FSK applies the data signal to the transmitter VFO, while AFSK transmits tones via phone. Diddling the frequency (FSK) changes the tone on the receiving end. Audio FSK feeds the audio tone into the microphone. From there, it is transmitted as SSB.

ALE stations establish contact by constantly scanning a list of frequencies, activating the radio when the designated call sign is received. ALE is Automatic Link Enable. The radio scans and automatically establishes contact.

E3 – RADIO WAVE PROPAGATION

E3A Electromagnetic Waves

The maximum separation between two stations communicating by EME is 12,000 miles, if the Moon is visible by both stations. *Hint: Forget the miles. The only way Earth-Moon-Earth works is if the Moon is visible by both stations. Common sense.*

Libration fading of an EME signal is a fluttery irregular fading. Libration is a perceived oscillating motion of two orbiting bodies. It would be irregular.

When scheduling an EME contact, the condition of least path loss is when the Moon is at perigee. Perigee is nearest to the Earth, so your signal travels a shorter distance. Apogee is afar.

Hepburn maps predict the probability of tropospheric propagation.

Tropospheric propagation often occurs along warm and cold fronts. The difference in temperature bends radio waves.

If DX signals become too weak to copy a few hours after sunset, switch to a lower frequency band. After dark, the lower frequencies come to life.

Atmospheric ducts capable of propagating microwave signals often form over bodies of water. *Hint: Microwaves bounce off water vapor.*

Meteor scatter is formed by free electrons in the E layer. *Cheat: Meteor, free, electrons = lots of Es.*

Meteor scatter is most suited to 28 MHz – 148 MHz. *Hint: Six meters is popular for meteor scatter,*

PROPAGATION

and this answer is the only one that includes 50 MHz (six meters). "Meteor" has six letters, and that helps me remember "six meters."

The atmospheric structure that can create a path for microwave propagation is temperature inversion. The signal bounces between layers of different temperatures.

The typical range for tropospheric propagation is 100 – 300 miles.

Auroral activity is caused by interaction in the E layer of charged particles from the Sun with the Earth's magnetic field. Hint: Aurora is associated with charged particles from the Sun. Don't bother with which layer.

The best mode for aurora propagation is CW. Hint: CW has a 10 dB advantage over SSB and is always the best mode for propagation compared to the other answers.

Circularly polarized electromagnetic waves are waves with a rotating electric field. Hint: Circles rotate.

E3B Transequatorial Propagation

Transequatorial propagation is between two mid-latitude points at approximately the same distance north and south of the magnetic equator. Hint: If it is _trans_equatorial, it must cross the equator.

The approximate maximum range for transequatorial propagation is 5,000 miles.

PROPAGATION

The best time for transequatorial propagation is afternoon or early evening. *Hint: Give the sun all day to charge up those electrons.*

The terms extraordinary and ordinary waves refer to independent waves created in the ionosphere that are elliptically polarized. *Cheat: Extraordinary and ordinary are independent.*

Linearly polarized waves that split into ordinary and extraordinary become elliptically polarized. *Hint: Look for "elliptically polarized" in answers about extraordinary waves.*

Long-path propagation is supported by 160 – 10 meters. *Hint: All HF bands support long path – going around the Earth the long way.*

The band that most frequently provides long-path propagation is 20 meters. *Hint: The band usually open for DX is 20 meters.*

The time of year sporadic E propagation is most likely to occur is around the solstices, especially the summer solstice. Sporadic E is a summer phenomenon.

Sporadic E propagation can occur any time of day. *Cheat: It is sporadic and can occur any time.*

The primary characteristic of chordal hop propagation is successive ionospheric reflections without an intermediate reflection from the ground. Chordal hop means the signal reflects inside the ionosphere and doesn't hit the Earth until it finally comes back down at the receiver.

Chordal hop is desirable because the signal experiences less loss compared to multi-hop

PROPAGATION

using Earth as a reflector. *Hint: Less loss is a good thing.*

E3C Radio Propagation

Ray tracing refers to modeling a radio wave's path through the ionosphere. *Hint: You trace the wave's path.*

A rising A or K index indicates increasing disruption of the geomagnetic field. *Hint: The index rises with geomagnetic disturbances.*

When the A or K index is elevated, the paths most likely to experience high levels of absorption are polar. *Hint: An elevated A or K index indicates a geomagnetic disturbance, and that would be strongest around the poles.*

The value of Bz represents the direction and strength of the interplanetary magnetic field. *Hint: B and z represent two things, direction and strength. The other answers are only one thing.*

The orientation of Bz that increases the likelihood incoming particles from the Sun will cause disturbed conditions is southward. *Hint: If the waves are headed southward (to the south), they will hit the North magnetic pole and cause disruptions.*

You learned on the Technician Class test that UHF/VHF radio waves can bend slightly over the horizon. **The radio horizon can exceed the geometric horizon by about 15% of the distance.** *Cheat: Not a lot, and this is the lowest answer.*

Solar flares are categorized by a letter. **The greatest solar flare intensity is Class X.** *Hint: X as in "eXtreme."*

PROPAGATION

Space weather also gets a letter. **The term G5 means an extreme geomagnetic storm.** *Hint: "G" as in "geomagnetic."*

The intensity of an X3 flare is 50% greater than an X2 flare. *Hint: 3 is 50% greater than 2.*

The 304A solar parameter measures UV emissions at 304 angstroms, correlated to the solar flux index. *Cheat: 304 A as in "Angstroms."*

As the frequency is increased, the maximum distance of ground wave propagation decreases. The ground attenuates higher frequencies more.

The type of polarization best for ground-wave propagation is vertical. *Hint: AM radio towers are vertical.*

The radio-path horizon distance exceeds the geometric horizon because of downward bending due to density variations in the atmosphere. *Hint: The air bends the wave over the horizon.*

A sudden rise in radio background noise might indicate a solar flare has occurred.

E4 – AMATEUR PRACTICES

E4A Test Equipment

The highest frequency signal that can be accurately displayed on a digital oscilloscope is limited by the sampling rate of the analog-to-digital converter. *Hint: The frequency is limited by the rate the oscilloscope can sample: a higher sampling rate = higher frequency.*

A spectrum analyzer would display RF amplitude and frequency. *Hint: It is analyzing the spectrum and shows the frequency and strength of the signal.*

To display spurious signals and/or intermodulation distortion products in an SSB transmitter, use a spectrum analyzer. *Hint: To display spurious signals, use a spectrum analyzer to see them.*

Calibrating an oscilloscope probe is called "compensating." **Compensation of an oscilloscope probe is typically done by displaying a square wave and adjusting the probe, so the horizontal lines are as flat as possible.** *Hint: A square wave display shows symmetrical flat lines.*

A prescaler on a frequency counter divides a higher-frequency signal so a low-frequency counter can display it. *Hint: It scales the signal before it is measured.*

The effect of aliasing on a digital oscilloscope caused by setting the time base too slow is a false, jittery low-frequency version of the signal is displayed. *Hint: Forget why it happens. Aliasing is another word for distortion, false display.*

AMATEUR PRACTICES

When using an oscilloscope probe it is good practice to keep the signal ground connection as short as possible. *Hint: A short ground lead has less effect on the measured circuit.*

The advantage of an antenna analyzer over an SWR bridge is antenna analyzers do not need an external RF source. *Hint: An antenna analyzer generates its own RF.*

A device that measures SWR is an antenna analyzer.

An antenna analyzer should be connected directly to the feed line. Antenna analyzers have a coax jack to plug in the coax.

To display multiple digital signal states simultaneously, use a logic analyzer. Signal states used in logic circuits are high/low or on/off.

E4B Measurements

The accuracy of a frequency counter is most affected by the time base accuracy. *Hint: The base must be accurate.*

The significance of voltmeter sensitivity expressed in ohms per volt is a full-scale reading of the voltmeter multiplied by its ohms per volt will indicate the input impedance. *Hint: Algebra. Volts x ohms/volt = ohms. Ohms indicate impedance. Cheat: The word "voltmeter" is in both the question and answer.*
The subscripts of S parameters represent the port or ports at which the measurements are made. *Cheat: S as in "portS."*[4]

AMATEUR PRACTICES

The S parameter equivalent to forward gain is S21. *Cheat: We looked forward to turning 21.*

The S parameter that represents return loss or SWR is S11. *Cheat: A 1:1 SWR is good.*

To calibrate a RF vector network analyzer, you use a short circuit, open circuit and 50-ohm loads. *Hint: You calibrate at the two extremes and the "normal" load of 50-ohm coax.*

A vector network analyzer, can measure:
- **Input impedance**
- **Output impedance**
- **Reflection coefficient**
- **All these choices are correct**

It analyzes the network parameters of an electrical circuit: in, out, and reflection.

The power absorbed by the load when the forward power is 100 watts and the reflected power is 25 watts is 75 watts. *Hint: Simple math. 100 watts went out, and 25 came back so, 75 watts must have stayed in the load.*

A relative measurement of the Q for a series-tuned circuit is the bandwidth of the circuit's frequency response. Higher "Q" means narrower bandwidth. *Hint: You measure Q by measuring the bandwidth at the frequency. Forget about "series tuned."*

---

[4] Sometimes there are random questions that mean nothing to you. The only way to recognize the correct answer is with a cheat. These cheats have nothing to do with a correct answer but will help you recognize one.

AMATEUR PRACTICES

The current reading on an RF ammeter in series with the antenna feed line will increase if there is more power going into the antenna. *Hint: More amps = more power.*

To measure intermodulation distortion (IMD) in an SSB transmitter, modulate with two non-harmonically related audio frequencies and observe the output on a spectrum analyzer. *Hint: You modulate with audio, and you measure distortion with a spectrum analyzer.*

E4C Receiver Performance

The effect of excessive phase noise in the local oscillator of a receiver is it can combine with strong signals on nearby frequencies to generate interference. *Hint: Excessive noise causes interference.*

The receiver circuits that can be effective in eliminating interference from strong out-of-band signals are a front-end filter or pre-selector. *Hint: The signals are out-of-band and need to be stopped at the front end before they get into the more sensitive internals of the receiver.*

A roofing filter is applied before the more sensitive IF circuits. **A narrow-band roofing filter affects receiver performance because it improves the dynamic range by attenuating strong signals near the receive frequency.** *Hint: Filters attenuate signals near the receive frequency.*

The term for the suppression in an FM receiver by another stronger signal on the same frequency is capture effect. *Hint: FM receivers capture the strongest signal.*

AMATEUR PRACTICES

The noise figure of a receiver is defined as the ratio in dB of the noise generated by the receiver to the theoretical minimum noise. *Hint: the question asks about the noise figure of a receiver. The answer mentions noise generated by the receiver. Again, the question gives away the answer.*

The value of -174 dBm/Hz noise floor represents the theoretical noise at the input of a perfect receiver at room temperature. *Cheat: Guaranteed to be the only answer on the test to contain "room temperature."*

A CW receiver with the AGC off has an equivalent input noise power density of -174 dBm/Hz. The level of unmodulated carrier input to this receiver that would yield an audio output SNR of 0dB in a 400 Hz noise bandwidth is -148 dBm. *Cheat: You've got to be kidding. Memorize -148 dBm*

The MDS of a receiver is the minimum discernible signal. It is a measure of receiver sensitivity.

An SDR is a software-defined radio. An analog-to-digital converter is the heart of an SDR. It converts the signals to digital ones and zeros. Then, computer power sorts them out. **An SDR receiver is overloaded when signals exceed the reference voltage of the analog-to-digital converter.** *Hint: Signals exceeding the reference voltage overload the receiver.*

Conventional super-heterodyne receivers convert the incoming signal to a fixed intermediate frequency (IF) and feed it through filters designed for that frequency. **A good reason for selecting a high frequency for the IF is it is easier for the front end circuitry to eliminate image responses.** Mixing "up" to a higher frequency puts the images farther away from the intermediate frequency.

AMATEUR PRACTICES

Mixing can produce images. If the intermediate frequency is 455 kHz, the converter will be 455 kHz above or below the tuned frequency. That can mix with other signals to produce an image on the tuned frequency. **A receiver tuned to 14.300 MHz with a 455 kHz IF frequency would also receive a signal at 15.210 MHz.** The local oscillator will be running at 14.755 MHz to generate a 455 kHz IF. But another signal at 15.210 would also mix with 14.755, resulting in 455 kHz out. *Hint: A shortcut is to double the IF and add it to the tuned frequency.*

Reciprocal mixing is local oscillator phase noise mixing with adjacent strong signals to create interference. *Hint: Local oscillator phase noise causes reciprocal mixing.*

The advantage of having a variety of receiver IF bandwidths to choose is receiver bandwidth can be set to match the modulation bandwidth. You can use a narrow filter for CW and a wider one for SSB.

An attenuator can be used to reduce receiver overload with little or no impact on the signal-to-noise ratio when atmospheric noise is greater than internally generated noise. You reduce noise to only the level generated in the receiver, so the signal stands out.

The largest effect on an SDR receiver's dynamic range is the analog-to-digital converter sample width. *Hint: Look for analog-to-digital converter in the answer. An SDR needs one.*

E4D Receiver Performance

The blocking dynamic range of a receiver is the difference in dB between the noise floor and the

AMATEUR PRACTICES

level of an incoming signal that will cause 1 dB of gain compression. *Hint: How strong a signal does it take to compress the gain 1 dB. Cheat: Recognize the answer with 1 dB.*

Two problems caused by poor dynamic range in a receiver are spurious signals caused by cross modulation of the desired signal and desensitization from strong adjacent signals. *Hint: Strong adjacent signals desensitize a receiver with poor dynamic range.*

Intermodulation interference can occur between two repeaters when the repeaters are in close proximity and the signals mix in the final amplifier of one or both. *Hint: Modulation is mixing.*

To reduce or eliminate intermodulation interference in a repeater, use a properly terminated circulator at the output of the transmitter. *Hint: To eliminate, terminate. A circulator is a one-way valve that shunts off signals coming down the transmission line and sends them to a dummy load (properly terminated).*

If a receiver tuned to 146.70 MHz and a nearby station transmits on 146.520 MHz, the transmitter frequencies which would produce an intermodulation-product signal in the receiver are 146.34 MHz and 146.61 MHz. *Hint/Cheat: Take the average of the two (146.70 + 146.52)/2 and look for the answer that has 146.61. You'll only solve one, but that is enough to choose the correct answer.*

The term for spurious signals generated by the combination of two or more signals in a non-linear device or circuit is intermodulation. *Hint: Intermodulation causes spurious signals.*

AMATEUR PRACTICES

Intermodulation in electronic circuits is caused by non-linear circuits or devices.

The term for a reduction in receiver sensitivity caused by a strong signal near the received frequency is desensitization.

A way to reduce desensitization is to decrease the RF bandwidth of the receiver. *Hint: Switch in a narrower filter to cut down the signal causing desensitization.*

The purpose of a preselector in a receiver is to increase rejection of signals outside the desired band. *Hint: It preselects signals.*

A third-order intercept level of 40 dBm means a pair of 40 dBm signals will generate a third-order intermodulation product with the same amplitude as the input signals. *Cheat: A pair are the same.*

Odd-order intermodulation products are of particular interest because the odd-order product of two signals which are in a band of interest will also likely be within the band.
Hint: You are interested because the products will be in the band you are working in.

E4E Noise Suppression and Grounding

One disadvantage of automatic notch-filtering on CW signals is it removes the CW signal at the same time. *Hint: The automatic filter removes all tones, including the CW tone.*
Noise that can be reduced with a digital signal processing noise filter includes:
- **Broadband white noise**
- **Ignition noise**
- **Power line noise**

AMATEUR PRACTICES

- **All these choices are correct**

Hint: Digital signal processing is effective against all sorts of noise.

A receiver noise blanker might be able to remove signals which appear across a wide bandwidth. A noise blanker is effective against pops from electric fences and spark plugs.

Conducted and radiated noise from an automobile alternator can be suppressed by connecting the radio's power leads directly to the battery, and by installing coaxial capacitors in line with the alternator leads. *Hint: Connecting directly to the battery is the best way to power a mobile rig. Look for that in the answer.*

Noise from an electric motor can be suppressed by installing a brute-force AC-line filter in series with the motor leads. *Hint: Suppress noise with brute-force.*

A nearby computer might cause unstable modulated or unmodulated signals at specific frequencies. *Hint: Computer circuits can generate "birdies" – signals heard at specific frequencies.*

Shielded cables can radiate or receive interference due to common mode currents on the shield and conductors. *Hint: The interference is common to the shield and conductors. Solve that by putting ferrite chokes on the cable.*

Current that flows equally on all conductors is called common-mode current. *Hint: It is common to all the conductors.*

An undesirable effect when using an IF noise blanker is signals may appear excessively wide even if they meet emission standards. *Cheat:*

The IF (intermediate frequency) is at radio frequencies. The other answers all relate to audio. If you can't recognize the correct answer, read the question and answers carefully for clues.

Loud roaring or buzzing that comes and goes could be:
- **Arcing contacts in a thermostatically controlled device.**
- **Defective doorbell or doorbell transformer.**
- **Malfunctioning illuminated advertising display.**
- **All these choices are correct.**

Hint: All operate intermittently are therefore cause noise that comes and goes.

If you hear combinations of local AM broadcast stations within MF or HF ham bands, it most likely is caused by nearby corroded metal joints mixing and re-radiating the broadcast signals.
Corroded metal joints can act as diodes and mix signals.

E5 – ELECTRICAL PRINCIPLES

E5A Resonance and Q

The voltage across reactance in series can be larger than the voltage applied due to resonance. *Those high voltages can cause arcing in your antenna tuner. Hint: Think of resonance as multiplying.*

Resonance in an electrical circuit is the frequency at which the capacitive reactance equals the inductive reactance. *When the two are equal, they cancel each other, and the circuit resonates.*

The magnitude of the impedance in a series RLC circuit at resonance is approximately equal to the circuit resistance. *Hint: "RLC" refers to resistor, inductor, and capacitor. If the inductor and capacitor cancel each other out, all that is left is resistance.*

The magnitude of impedance with a resistor, inductor, and capacitor all in parallel at resonance is equal to the circuit resistance. *Hint: In series, or parallel, at resonance, the answer is the same. All that is left is the resistance.*

"Q" stands for reactive quotient and is a measure of selectivity. A high Q circuit is very selective and narrow banded. **The effect of increasing the Q in an impedance matching network is the matching bandwidth is decreased.**

The magnitude of the circulating current in a parallel LC circuit at resonance is maximum. *Hint: The <u>circulating</u> current in the circuit is maximum. The inductive and capacitive reactance are*

ELECTRICAL PRINCIPLES

in parallel, and the current gets transferred back and forth between them (circulating).

The magnitude of the current at the input of a parallel RLC circuit at resonance is minimum. Hint: *The circulating current stays in the circuit and doesn't draw current from the supply. The glass is full. The input current is at a minimum.*

The phase relationship between the voltage and current in a series resonate circuit is the voltage and current are in phase. Hint: *In a resonant circuit, inductance and capacitance cancel each other and voltage and current are in phase.*

The Q of an RLC parallel resonant circuit is calculated by resistance divided by reactance.

The Q of an RLC series resonant circuit is calculated by the reactance divided by the resistance.

Memory point:
Q in Parallel = Resistance/reactance
Q in Series = Reactance/resistance

The half-power bandwidth of a resonate circuit that has a resonant frequency of 3.7 MHz, and a Q of 118 is 31.4 kHz. Half-power bandwidth is frequency divided by Q.
Solve: 3.7 MHz/118 = .3135 MHz is 31.35 kHz and 31.4 kHz is the closest. The answers are far enough apart you can get sloppy with your decimal points. Look for the solution with the correct integers.

The half-power bandwidth of a resonant circuit that has a resonate frequency of 7.1 MHz, and a Q of 150 is 47.3 kHz.
Solve: 7.1MHz/150 = .0473 MHz = 47.3 kHz.

Extra Class – The Easy Way

ELECTRICAL PRINCIPLES

Increasing Q in a series resonant circuit increases internal voltages. *Hint: Increasing Q in a series circuit increases reactance, which increases voltages. In a series circuit, Q=reactance/resistance.*

The resonate frequency of a series RLC circuit if R is 22 ohms, L is 50 microhenrys and C is 40 picofarads is 3.56 MHz.

The resonate frequency of a parallel RLC circuit if the R is 33 ohms, L is 50 microhenrys and C is 10 picofarads is 7.12 MHz.

The formula for resonant frequency is F=1/(2π X √LC). If that math is too much for you, as it is for me, there are online calculators, so I don't feel bad about giving you this cheat: *The answer is either in the 80 or 40-meter ham band.*

To increase Q in inductors and capacitors, lower losses. *Hint: The lossy resistance is what makes the circuit less sharp (lower Q).*

E5B Time and Phase

The term for the time required for the capacitor in an RC circuit to be charged to 63.2% of the applied voltage or to discharge to 36,8% of its initial voltage is "one time constant."

The time constant of a circuit having two 220-microfarad capacitors and two 1-megaohm resistors in parallel is 220 seconds. : The formula is resistance times capacitance. *Solve: Two 220-microfarad capacitors in parallel equal 440 microfarads. Two 1-megaohm resistors in parallel equal .5 megaohms. 440 x .5 = 220. Cheat: Remember the answer is the value of the capacitor.*

ELECTRICAL PRINCIPLES

Susceptance is the imaginary part of admittance. Susceptance is the measure of how much a circuit is *susceptible* to conducting a changing current.

If the magnitude of a pure reactance is converted to susceptance it becomes the reciprocal. Resistance is opposition to DC current. Its opposite is conductance. Reactance is a measure of how a circuit reacts to AC current and susceptance is the reciprocal of reactance (1/X).

The letter B is used to represent susceptance.

Impedance is opposition to AC current. **Admittance is the inverse of impedance.** *Hint: Impedance impedes and admittance admits.*

Impedance in polar form converted to admittance is the reciprocal of the magnitude and change the sign of the angle. *Hint: Admittance is the inverse. Take the reciprocal of the magnitude and change the sign of the angle.*

Impedance makes the voltage or current lead one another. In an inductive circuit, voltage leads the current. In a capacitive circuit, voltage lags the current. Remember, "ELI the ICEman." E is voltage, L is inductance, I is current, C is capacitive.

The relationship between current through a capacitor and the voltage across it are that the current leads the voltage by 90 degrees. *Hint: Current leads voltage in a capacitor. ELI the ICEman told me so.*

The relationship between the current through and the voltage across an inductor is voltage leads the current by 90 degrees. *Hint: Voltage leads current in an inductor says ELI.*

ELECTRICAL PRINCIPLES

The phase angle between the voltage across and the current through a series RLC circuit if the XC is 500 ohms, the R is 1 kilohm ohms and XL is 250 ohms is 14 degrees with the voltage lagging the current. Hint: We know the voltage is lagging the current because the XC is greater than the XL. That eliminates two answers. The rest of the math is staggering. Cheat: The answer to all questions is 14 degrees. The circuit is more capacitive, so the ICEman says voltage is lagging.

The phase angle between the voltage across and the current through a series RLC circuit if the XC is 100 ohms, the R is 100 ohms and XL is 75 ohms is 14 degrees with the voltage lagging the current. Hint: We know the voltage is lagging the current because the XC is greater than the XL. That eliminates two answers. Cheat: The answer to both questions is 14.

The phase angle between the voltage across and the current through a series RLC circuit if XC is 25 ohms, R is 100 ohms, and XL is 50 ohms is 14 degrees with the voltage leading the current. Hint: Same 14 degrees, but the circuit is inductive so voltage leads current. ELI.

E5C Coordinate Systems

Impedances are described in polar coordinates by phase angle and magnitude.

The coordinate system used to display the phase angle of a circuit containing resistance, inductance and/or capacitive reactance is polar coordinates. Hint: Phase angle is polar coordinates.

Memory point: Phase angle = polar coordinates.

ELECTRICAL PRINCIPLES

The coordinate system used to display resistive, inductive and/or capacitive reactance components of impedance is rectangular coordinates. *Hint: If the system does not mention phase angle, it is rectangular coordinates.* Figure E5-1 is an example.

A capacitive reactance in rectangular notation is –jx. *Hint: Look at ELI the ICEman's voltage. This is a capacitive circuit, so the voltage is behind, -j.*

In polar coordinates, an inductive reactance has a positive phase angle. Hint: *This is an inductive circuit so the voltage is ahead of current (positive angle, +j).*

The diagram used to show the phase relationship between impedances at a given frequency is called a phasor diagram. *Cheat: If you see a Star Trek weapon in the answer, it is correct.*

An impedance of 50-j25 represents 50 ohms resistance with 25 ohms capacitive reactance. The first number is ohms resistance. The second is reactance. Since j is negative, we know the voltage is lagging, and the circuit must be capacitive.

When using rectangular coordinates to graph the impedance of a circuit, the horizontal axis is the resistive component. *Cheat: Horizontal sounds like reziztive.*

The following questions refer to figure E5-1 on the next page:
The point representing a 400-ohm resistor and a 38 picofarad capacitor at 14 MHz is point 4.
Solve: We know the answer lies on the +400-ohm line of the horizontal axis (resistance). We know the circuit is capacitive, so it will have a minus sign and be

Extra Class – The Easy Way   Page 47

ELECTRICAL PRINCIPLES

on the lower quadrant on the vertical axis. Point 4 it is, by default.

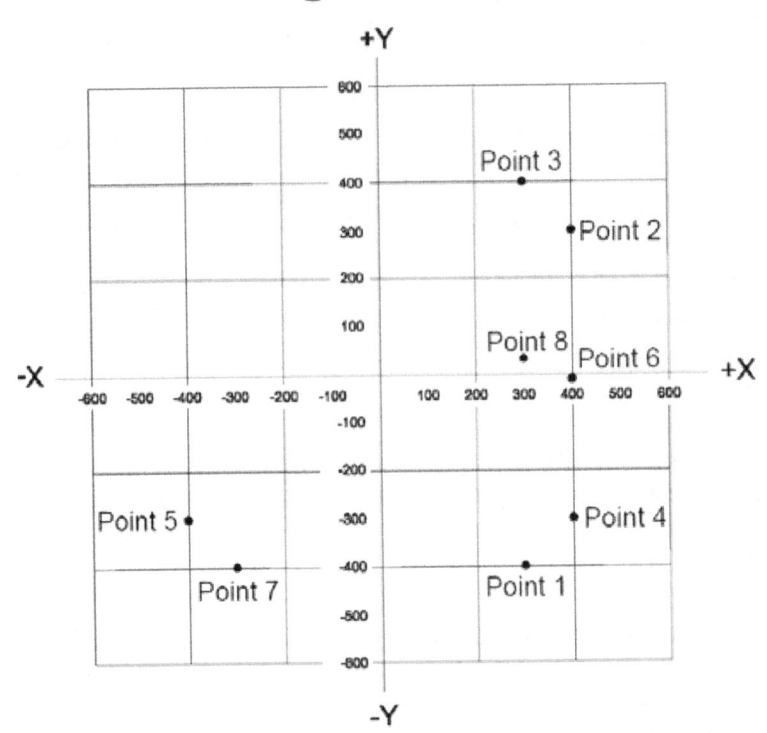

Figure E5-1

[Graph showing an X-Y coordinate plane with axes from -600 to +600 on both X and Y. Points plotted:
- Point 1: approximately (300, -400)
- Point 2: approximately (400, 300)
- Point 3: approximately (300, 395)
- Point 4: approximately (500, -300)
- Point 5: approximately (-500, -300)
- Point 6: approximately (500, 100)
- Point 7: approximately (-300, -400)
- Point 8: approximately (300, 100)]

The point representing a circuit has a 300-ohm resistor, and an 18 millihenry inductor at 3.5 MHz is point 3. *Solve:* Inductive reactance is 2π times frequency times inductance. 2x3.14x3.5x18= 395. The circuit is inductive. Look up on the positive Y side, starting at the 300-ohm point on the X axis. Point 3 is the correct answer, where 300 and 395 cross.

The point representing a circuit that has a 300-ohm resistor and a 19 picofarad capacitor at 21.200 MHz is point 1. *Solve:* On a vertical line off the 300-ohm horizontal axis, and the circuit is capacitive, so the answer is lower (-Y) quadrant. Point 1 by default.

Cheat: 300-ohm resistor is either point 1 or 3. If the circuit is inductive, Y is positive, point 3. If the circuit is capacitive, Y is negative, point 1.

E5D AC and RF Energy

The result of skin effect is, as frequency increases, RF flows in a thinner layer of the conductor, closer to the surface.

It is important to keep lead lengths short for components in circuits for VHF and above to avoid unwanted inductive reactance.

Short connections are used at microwave frequencies to avoid phase shift along the connection. *Hint: Another way to say "avoid unwanted reactance."*

Microstrip is precision printed circuit conductors above a ground plane that provide constant impedance interconnects at microwave frequencies. *Hint: Simplify it. "Constant impedance" is a good thing.*

The direction of a magnetic field about a conductor in relation to the direction of the electron flow is oriented in a circle around the conductor. *Hint: Forget the current flow, the magnetic field is in a circle around the conductor.*

The true power in an AC circuit where the voltage and current are out of phase can be determined by multiplying the apparent power times the power factor. When the current and voltage are out of phase, the inductor and capacitor are feeding each other. Some power recirculates, so the apparent power is higher. Multiply apparent power by the power factor to see what the circuit consumed.

ELECTRICAL PRINCIPLES

The power factor of an R-L circuit having a 60-degree phase angle between the voltage and the current is .5 *Solve: The answer is the cosine of the phase angle. The cosine of 60 degrees is .5. You will have to bring a scientific calculator to the test or memorize three values.*

The power factor of an R-L circuit with a 30-degree phase angle is .866. *Solve. The cosine of 30 degrees is .866.*

The power factor of an R-L circuit with a 45-degree phase angle is .707. *Solve: the cosine of 45 degrees is .707.*

Memory point: Higher angle = lower cosine
Cosine 60 degrees is .5
Cosine 45 degrees is .707
Cosine 30 degrees is .866

Watts consumed in a circuit having a power factor of .71 with an apparent power of 500VA would be 355W. *Solve: 500 x .71 = 355 watts.*

If a circuit has a power factor of .6 and the input is 200VAC at 5 amperes, 600 watts are consumed. *Solve: First, convert the volts and amps to watts: P=EI, P= 200 x 5 = 1000. Multiply by the power factor of .6 = 600 watts.*

If a circuit has a power factor of .2 and the input is 100 VAC at 4 amperes, the watts consumed are 80 watts. *Solve: Convert to watts: 100 x 4 = 400 watts. Times the power factor of .2 = 80 watts.*

Reactive power is wattless and non-productive power. It is the capacitor and inductor feeding each other.

ELECTRICAL PRINCIPLES

The reactive power in an AC circuit that has ideal inductors and capacitors is repeatedly exchanged between the associated magnetic and electric fields but is not dissipated. *Hint: If the components are ideal, energy is not dissipated. Look for "not dissipated" in the correct answer.*

The power consumed in a circuit consisting of a 100-ohm resistor in series with a 100-ohm inductive reactance drawing 1 ampere is 100 watts. *Foul! This is a trick question. It assumes no loss in the inductor. How much power is consumed by the resistor portion of the circuit? You are supposed to ignore the power consumed in the inductor. The answer is $P=I^2R$. I^2 is 1; times the R (100) is 100 watts.*

E6 – CIRCUIT COMPONENTS

E6A Semiconductors

Gallium arsenide is used as a semiconductor material in microwave circuits.

The semiconductor material that contains excess free electrons is N-type. *Hint: Excess free electrons would be **N**egatively charged.*

A PN-junction does not conduct current when reverse biased because holes in the P-type material and electrons in the N-type material are separated by the applied voltage widening the depletion region. *Hint: Reverse bias separates and widens the depletion region.*

The name given to an impurity that adds holes to a semiconductor crystal structure is acceptor impurity. *Hint: Holes accept electrons.*

The DC input impedance at the gate of a field-effect transistor is higher than a bipolar transistor. FETs have high input impedance.

The beta of a bipolar transistor is the change in collector current with regard to base current. The change between the collector and base. *Cheat: Beta and base current.*

A silicon NPN transistor is biased on when the base-to-emitter voltage is approximately .6 - .7 volts. *Hint: It is biased by voltage, eliminating 2 of the answers. The bias voltage is small, eliminating the other answer.*

The term indicating the frequency at which a grounded-base current gain of a transistor has

CIRCUIT COMPONENTS

decreased to .7 of the gain obtainable at 1 kHz is called the alpha cutoff frequency. *Hint: Gain (alpha) is being cut off (decreased).*

A depletion-mode FET is an FET that exhibits a current flow between source and drain when no gate voltage is applied. *Hint Current flow with no gate voltage would deplete.*

In Figure E6-1, the symbol for an N-channel dual-gate MOSFET is number 4. *Hint: It is dual gate and only two of the figures show two gates (G1 and G2). It is N channel so the arrow is pointed in. Note: This is the opposite of bipolar transistors where the N type has the arrow not pointed in.*

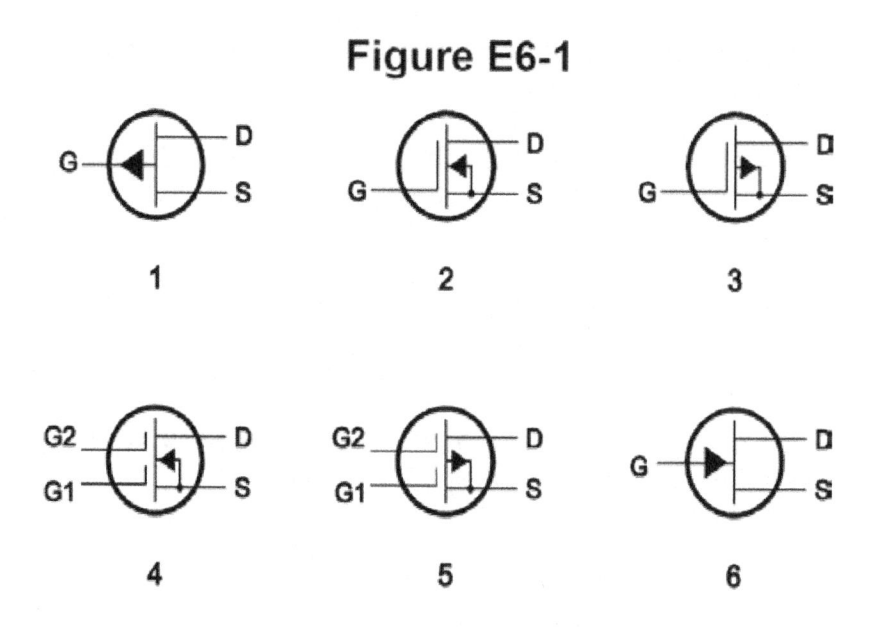

Figure E6-1

The symbol for a P-channel junction FET is number 1. *Hint: A P-channel FET has the arrow pointed out. These are the only two symbols you need to know from Figure E6-1.*

CIRCUIT COMPONENTS

Many MOSFET devices have internally connected Zener diodes on the gates to reduce the chance of static damage to the gate.

E6B Diodes

The most useful characteristic of a Zener diode is **constant voltage drop under conditions of varying current.**

The important characteristic of a Schottky diode is **less forward voltage drop.**
Cheat: Look for "voltage drop" in both answers.

A common use of a Schottky diode is as a **VHF/UHF mixer or detector.**

A type of Schottky barrier diode is **metal semi-conductor junction.**

The bias to light an LED is forward bias. *Hint: It is conducting – moving forward. This is the only symbol you need to know on Figure E6-2.*

The semiconductor device designed for use as a voltage-controlled capacitor is a **varactor diode.** *Hint: The voltage varies the capacitance.*

The characteristic of a PIN diode that makes it useful as an RF switch is **low junction capacitance.** *Hint: If it had high capacitance, the charge would hinder the switching ability.*

To control the attenuation of RF signals by a PIN diode, use **forward DC bias current.** *Hint: Give the signals a boost with forward bias.*

A common use for a point-contact diode is as an **RF detector.** *Hint: An RF detector helps you make a contact.*

CIRCUIT COMPONENTS

When a junction diode fails due to excessive current, the mechanism is excessive junction temperature. *Hint: Excessive current causes excessive heat, which causes a failure.*

The schematic symbol for a light-emitting diode in Figure E6-2 is number 5. *Hint: The arrows are light coming off the device. This is the only symbol you need to remember from E6-2.*

Figure E6-2

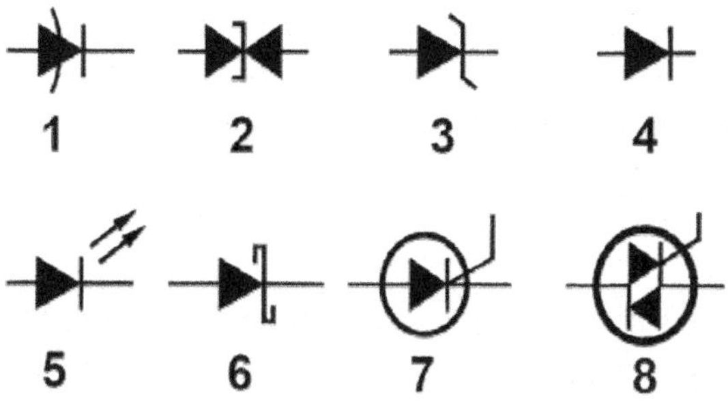

E6C Digital ICs

A comparator is a device that compares two voltages or currents and outputs a digital signal indicating which is larger. **When the level of a comparator's input signal crosses the threshold, the comparator changes its output state.**

The function of hysteresis in a comparator is to prevent input noise from causing unstable output signals. Hysteresis is a voltage lag between active and inactive states. A 12-volt relay clicks on at 11 volts but doesn't click off until voltage drops to 9 volts. Small variations in voltage don't trigger it.

Tri-state logic is logic devices with 0,1 and high impedance output states. *Hint: Tri-state is three output states: 0, 1 and high.*

CIRCUIT COMPONENTS

BiCMOS logic is an integrated circuit logic family using both bipolar and CMOS transistors (Complementary Metal Oxide Semiconductor). **An advantage of BiCMOS logic is it has the high input impedance of CMOS and the low output impedance of bipolar transistors.** *Hint: High input impedance is a good thing because it doesn't load down the circuit.*

The advantage of CMOS logic devices over TTL devices is lower power consumption.

CMOS digital integrated circuits have high immunity to noise on the input signal or power supply because the input switching threshold is about one-half the power supply voltage. *Hint: It switches long before power supply voltage variations have any effect.*

A pull-up or pull-down resistor is connected to the positive or negative supply line used to establish a voltage when an input or output is an open circuit. *Hint: Pull up or pull down means it could be positive or negative. The resistor sets the bias.*

A Programmable Logic Device (PLD) is a programmable collection of logic gates and circuits in a single integrated circuit.

In Figure E6-3, the schematic symbol for a NAND gate is number 2.

A NAND gate produces logic 0 at its output only when all inputs are logic 1. *Hint: N as in negative or reversed. The little circle on the output means the output is reversed. The two inputs are on a straight vertical line; they are the same.*

CIRCUIT COMPONENTS

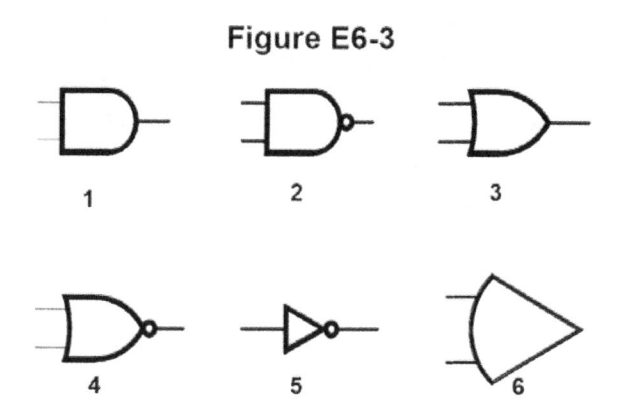

Figure E6-3

Figure E6-3 has symbols for logic circuits. You only need to recognize 2, 4 and 5.

A NOR gate produces a logic 0 if any inputs are logic 1. *Hint: The N means it produces an opposite logic. The OR means either input could be 1.*
The symbol for a NOR gate is number 4. *Hint: The inputs are on a curved line (OR), and the output has the little circle indicating reversed output.*

The symbol for the NOT operation (inverter) is number 5. *Hint: One input and one output are all that is required to invert.*

E6D Torroidal and Solenoidal Inductors;

Inductor saturation is when the ability of the core to store magnetic energy has been exceeded. *Hint: The core is saturated.*

Core saturation should be avoided because harmonics and distortion could result. *Hint: The core can't store any more magnetic energy, so there is distortion.*

CIRCUIT COMPONENTS

The equivalent circuit of a quartz crystal is motional capacitance, motional inductance and loss resistance in series, all in parallel with a shunt capacitor representing electrode and stray capacitance. *Cheat: Horrors! You have got to be kidding. Just look for the answer with "shunt" in it and bypass all this.*

An aspect of the piezoelectric effect is mechanical deformation of material by the application of a voltage. *Hint: We usually think of it the other way around, a voltage generated by pressing the material; the way a grill lighter sparks.*

The materials commonly used as a core in an inductor are ferrite and brass.

The material that decreases inductance when inserted into a coil is brass. *Hint: Brass is not magnetic.*

A reason to use ferrite cores instead of powdered-iron in an inductor is ferrite toroids require fewer turns to produce a given inductance value. *Hint: Ferrite is more efficient*

A reason to use powdered-iron instead of ferrite is the powdered-iron cores maintain their characteristics at higher currents. *Hint: Iron can handle higher current because iron can withstand heat better.*

The material property that determines the inductance of a toroidal inductor is its permeability. *Permeability determines how the magnetic flux reacts. Think of it as a measure of the efficiency of the core.*

The current in the primary winding of a transformer with no load attached is called

CIRCUIT COMPONENTS

magnetizing current. *Hint: It is just enough to magnetize the core but no more because there is no load to draw current.*

The devices commonly used as VHF and UHF parasitic suppressors at the input and output terminals of a transistor HF amplifier are ferrite beads. *Hint: Ferrite beads are toroids. The beads act as chokes.*

The primary advantage of using a toroidal core over a solenoidal core is the toroidal core confines most of the magnetic field within the core material. A solenoidal core is a bar of ferrite with the wire wrapped around. The circular toroid keeps the magnetic field contained.

The primary cause of inductor self-resonance is inter-turn capacitance. The turns of wire interact as capacitors that form a resonant circuit with the inductor.

E6E Analog ICs; MMICs; IC Packaging

Gallium arsenide (GaAs) is useful for semiconductor devices operating at UHF and higher frequencies because it has higher electron mobility. *Hint: Electron mobility is a strange enough term it should stick with you.*

An example of a through-hole device is a DIP. Dual Inline Package is a type of computer chip. It has leads that are soldered through holes in the circuit board.

MMIC means Monolithic Microwave Integrated Circuit. **The material likely to provide the highest frequency of operation when used in MMICS is gallium nitride.** *Hint: Gallium is used at high*

CIRCUIT COMPONENTS

frequencies. Gallium is also the answer to another question.

The most common input and output impedance of a circuit that use MMICs is 50 ohms. Hint: Just like our coax.

The noise figure typical for a low-noise UHF preamplifier is 2 dB. Hint: The pre-amplifier will add some noise, but not a lot. The other answers are either negative or too high.

The characteristic of MMIC that makes it popular for VHF through microwave circuits is controlled gain, low noise figure, and constant input and output impedance over the specified frequency range. Hint: You don't need to remember all of that. Just pick "low noise."

The transmission line used for connections to MMICs is microstrip. Hint: Microwave = microstrip.

The component package most suitable for use above the HF range are surface mount. Hint: Surface mount has the shortest leads (none), and high frequencies require short leads to reduce stray reactance.

The advantage of surface-mount technology over through-hole components is:
- **Smaller circuit area**
- **Shorter circuit-board traces**
- **Components have less parasitic inductance and capacitance**
- **All these choices are correct**

Hint: Recognize two, and you know the answer is all.

CIRCUIT COMPONENTS

A characteristic of DIP packaging is two rows of connecting pins placed on opposite sides of the package. *Hint: Dual Inline Package.*

DIP through-hole package ICs are not typically used at UHF and higher frequencies because of excessive lead length.

E6F Optical Components

Electrons absorb the energy from light falling on a photovoltaic cell.

When light shines on a photoconductive material the conductivity increases. *Hint: Photoconductive.*

The most common configuration of an optoisolator or optocoupler is an LED and a phototransistor. Two circuits are isolated (not electrically connected), but signals pass via light.

The photovoltaic effect is the conversion of light to electrical energy.

An optical shaft encoder detects rotation of a control by interrupting a light source with a patterned wheel. *Hint: It senses light pulses when the shaft turns.*

A material most commonly used to create photoconductive devices is a crystalline semiconductor.

A solid-state relay is a device that uses semiconductors to implement the functions of an electromechanical relay. *Hint: A solid-state relay is a relay made with solid-state devices (semiconductors).*

CIRCUIT COMPONENTS

An optoisolator is often used in conjunction with solid state circuits when switching 120VAC because optoisolators provide a high degree of electrical isolation between the control circuit and the circuit being switched. *Hint: Optoisolators provide isolation. Look for that word in the answer.*

The efficiency of a photovoltaic cell is the relative fraction of light that is converted to current. *Hint: The more efficient, the more current it produces for a given amount of light.*

The most common type of photovoltaic cell used for electrical power is silicon. *Hint: Solar panels are made of silicon.*

The approximate voltage produced by a fully illuminated silicon photovoltaic cell is 0.5 V. Each cell produces 0.5 volts. Connect the cells in series to output a higher voltage.

E7 – PRACTICAL CIRCUITS

E7A Digital Circuits

A bi-stable circuit is a flip-flop. *Hint: "Bi" means it has two states, flip and flop.*

The function of a decade counter digital IC is to produce one output signal for every ten input pulses. *Hint: It counts decades or tens.*

The circuit that can divide the frequency of a pulse train by 2 is a flip-flop.

To divide a signal frequency by 4 requires 2 flip-flops. *Hint: 2 X 2.*

A circuit that continuously alternates between two states without an external clock is an astable multivibrator. *Hint: Astable means it alternates continuously. It is never stable.*

A monostable multivibrator switches momentarily to the opposite binary state and then returns to its original state after a set time. *Hint: Monostable means it returns to one state.*

A NAND gate produces a logic 0 when all inputs are logic 1. *Hint: N means it produces an opposite logic. AND means both inputs are the same.*

An OR gate produces a logic 1 if any or all inputs are logic 1. *Hint: There is no N in front, so output equals input. OR means either input could be 1.*

A NOR gate produces a logic 0 if only one input is logic 1. *Hint: The N means it produces an opposite logic. The OR means either input could be 1.*

PRACTICAL CIRCUITS

A truth table is a list of inputs and corresponding outputs for a digital device. *Hint: It is a table showing the various outcomes.*

The logic that defines "1" as a high voltage is positive logic. *Hint: 1 is a positive number.*

E7B Amplifiers

Amplifiers can operate over all or part of the 360-degree signal cycle. Their Class describes how much of the cycle conducts.

A push-pull Class AB amplifier conducts more than 180 degrees but less than 360 degrees.

A Class D amplifier uses switching technology to achieve high efficiency.

The components of a class D amplifier include a low pass filter to remove switching signal components. *Hint: Switching circuits can generate spurious signals and the low pass filter removes them.*

The load line of a Class A common emitter amplifier would normally be biased half-way between saturation and cutoff. *Hint: Class A runs all through the 360-degree cycle, so it would have to be biased right in the middle.*

To prevent unwanted oscillations in an RF power amplifier, install parasite suppressors and/or neutralize the stage. *Hint: Suppressors suppress unwanted oscillations. Neutralization is introducing some negative feedback to cancel the oscillations.*

An RF power amplifier can be neutralized by feeding in a 180-degree out-of-phase portion of the output back to the input.

PRACTICAL CIRCUITS

An amplifier that reduces or eliminates even-order harmonics is a push-pull design.

If a Class C amplifier is used to amplify SSB, you would experience signal distortion and excessive bandwidth. Class C operates less than 50% of the time. Class C is not linear and is suited for CW but not SSB.

When tuning a vacuum tube RF power amplifier that uses a Pi-network output circuit, adjust the tuning capacitor for minimum plate current and the loading capacitor for maximum permissible plate current. *Hint: Tune the plate for minimum and load for maximum.*

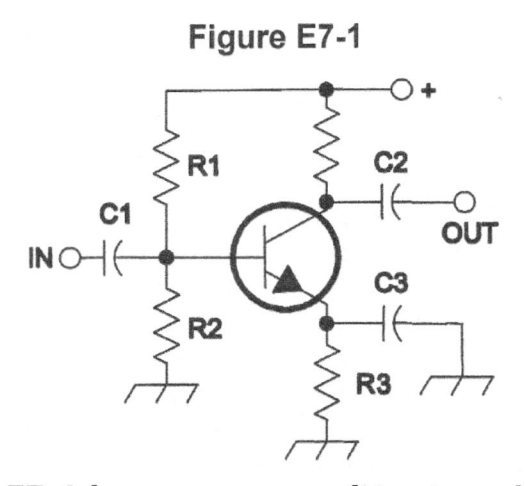

Figure E7-1

Figure E7-1 is a common emitter transistor amplifier.

There are three questions related to figure E7-1 and you only need to identify it as "common emitter" and know the purpose of R1/R2 and R3. **In Figure E7-1 the purpose of R1 and R2 is fixed bias.** They set the voltage on the base of the transistor. They form a voltage divider between the + and ground.

PRACTICAL CIRCUITS

The purpose of R3 is self bias. It is biasing the whole transistor. Amplifiers operate using signal inputs, which vary from positive to negative. Biasing establishes the correct operating point of the transistor and, if done correctly, reduces distortion.

An emitter follower (or common collector) amplifier is an amplifier with low impedance output that follows the base input voltage. *Hint: An emitter follower follows.*

Switching amplifiers are more efficient than linear amplifiers because the power transistor is at saturation or cutoff most of the time. *Hint: Another way of saying it only conducts for a short part of the cycle.*

To prevent thermal runaway in a bipolar transistor, use a resistor in series with the emitter. That restricts the amount of current that can flow and overheat the transistor.

The effect of intermodulation products in a linear power amplifier are transmission of spurious signals. *Hint: Intermodulation is distortion and can lead to spurious signals.*

Odd-order rather than even-order intermodulation distortion products are of greater concern in linear power amplifiers because they are relatively close in frequency to the desired signal. *Hint: Being close in frequency is a problem.*

A characteristic of a grounded-grid amplifier is low input impedance. *Hint: If the grid is grounded, it is at low impedance.*

PRACTICAL CIRCUITS

E7C Filters and Matching Networks

Capacitors and inductors in a low-pass filter Pi-network are arranged with a capacitor connected between the input and ground, another capacitor between the output and ground and an inductor is connected between the input and output. *Hint: Another overly-complicated answer. The filter is to pass low frequencies, so it must shunt off high frequencies. The way to do that is with capacitors to ground from both the input and output. Look for that as part of the answer.*

A property of a T-network with series capacitors and a parallel shunt inductor is a high-pass filter. *Hint: Series capacitors pass high frequencies so it is a high pass filter. The opposite of the above.*

A Pi-L network has an advantage over a Pi-network because it offers greater harmonic suppression. *Hint: The extra component (L) offers more suppression.*

An impedance matching circuit transforms a complex impedance to a resistive impedance by canceling the reactive part of the impedance and changing the resistive part to a desired value. *Hint: Impedance matching changes to a desired value. Look for "a desired value" in the answer.*

The filter with a ripple in the passband and a sharp cutoff is a Chebyshev filter. *Cheat: Recognize the Russian name. It is only in one answer.*

The distinguishing features of an elliptical filter are extremely sharp cutoff with one or more notches in the stop band. *Cheat: "Extremely sharp cutoff" is good in a filter and the answer.*

PRACTICAL CIRCUITS

A Pi-L network used for matching a vacuum tube final amplifier to a 50-ohm unbalanced output is a Pi network with an additional series inductor on the output. *Hint: Pi-L adds an L (inductor) to a Pi network. That is all you need to know.*

The factor having the greatest effect on the bandwidth and response shape of a crystal ladder filter is the relative frequencies of the individual crystals. *Hint: The crystal frequencies determine the shape of the filter.*

A crystal lattice filter is a filter with narrow bandwidth and steep skirts made using quartz crystals. Several crystals are connected in a lattice making the filter narrower and sharper. *Hint: The crystals aren't lattice-shaped; the circuit is.*

The best choice of a filter for a 2-meter repeater duplexer would be a cavity filter. Cavity filters are the basic circuitry behind a duplexer and are sharply tuned resonant circuits that allow only certain frequencies to pass. A duplexer isolates the receiver from the transmitter when they both use the same antenna.

The term used to describe a filter's ability to reject adjacent signals is shape factor. A narrow shape factor has steeper sides.

An advantage of a Pi-matching network over an L-matching network consisting of a single inductor and capacitor is the Q of Pi networks can be controlled. *Hint: A Pi-matching network has three components. An L-match has only two. A single inductor and capacitor mean you can't change both the input and output.*

PRACTICAL CIRCUITS

E7D Power Supplies

A linear electronic voltage regulator works by the conduction of a control element varied to maintain a constant output voltage. *Hint: It is a regulator, so it regulates to "maintain a constant output voltage."*

One characteristic of a switching electronic voltage regulator is the duty cycle is changed to produce a constant average output voltage. *Hint: Switching is changing the duty cycle to produce a constant average.*

The device typically used as a stable reference voltage in a linear voltage regulator is a Zener diode. Zener diodes are voltage regulators.

The linear voltage regulator that makes the most efficient use of the primary power source is a series regulator. *Hint: "Regulator" in the question and the answer. Because it is in series, power passes to the load. A shunt would dump some to ground.*

The linear voltage regulator that places a constant load on the unregulated voltage source is a shunt regulator. *Hint: "Regulator" in the question and answer. The shunt keeps a constant load by dumping excess to ground.*

There are only three possible questions related to Figure E7-2 on the following page. **The circuit shown in Figure E7-2 is a linear voltage regulator.**

PRACTICAL CIRCUITS

Figure E7-2

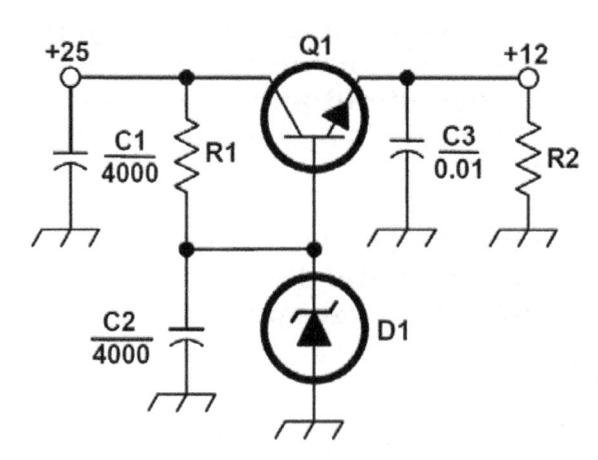

The purpose of Q1 in Figure E7-2 is to control the current supplied to the load. *Hint: The current is passing through Q1, a big power transistor, instead of the regulator. Q1 is between the input and output.*

The function of a pass transistor in a linear voltage regulator is to maintain a nearly constant output voltage over a wide range of load current. *Hint: Q1 is a pass transistor.*

The purpose of C2 in Figure E7-2 is to bypass rectifier output ripple around D1. *Hint: Capacitors are often used to bypass ripple (hum).*

The reason to use a charge controller with a solar power system is to prevent battery damage due to overcharge. *Hint: A charge controller controls the amount of charge. Overcharge will damage the battery.*

The reason a high-frequency switching type high voltage power supply can be both less expensive and lighter in weight than a conventional power

PRACTICAL CIRCUITS

supply is the high-frequency inverter design uses much smaller transformers and filter components. *Hint: "High frequency" in both the question and answer.*

The drop-out voltage of an analog voltage regulator is the minimum input-to-output voltage required to maintain regulation. *Hint: Without the minimum input-to-output voltage, it drops out and stops regulating.*

The equation for calculating power dissipation by a series linear voltage regulator is the voltage difference from input to output multiplied by output current. *Hint: You are solving for dissipation, the power kept in the regulator, and that depends on how much of a difference there is between input and output.*

The purpose of connecting equal-value resistors across power supply filter capacitors connected in series is to:
- **Equalize the voltage across each capacitor**
- **Discharge the capacitors when voltage is removed**
- **Provide a minimum load to the supply**
- **All these choices are correct**

The purpose of a "step-start" circuit in a high-voltage power supply is to allow the filter capacitors to charge gradually. *Hint: It starts in steps to avoid a possibly-damaging voltage surge.*

E7E Modulating and Demodulating

FM phone emissions can be generated by a reactance modulator on the oscillator. *Hint: It is FM, so you modulate the oscillator, and a reactance modulator is one way to do that.*

PRACTICAL CIRCUITS

The function of a reactance modulator is to produce FM using electrically variable inductance or capacitance. *Hint: "Reactance modulator," tells you it is FM. You vary the frequency with reactance from inductors and capacitors.*

A frequency discriminator stage in an FM receiver is a circuit for detecting FM signals.

One way to generate a single-sideband phone signal is using a balanced modulator followed by a filter. *Hint: You filter off the other sideband.*

The circuit added to an FM transmitter to boost the higher audio frequencies is a pre-emphasis network. *Hint: It emphasizes the higher audio frequencies.*

De-emphasis is used in FM communications receivers for compatibility with transmitters using phase modulation. It undoes the emphasis added by a pre-emphasis circuit and thereby makes the signal compatible.

The term "baseband" means the frequency range occupied by the message signal prior to modulation. *Hint: The base frequency range before modulation.*

The principal frequencies that appear at the output of a mixer circuit are the two input frequencies along with their sum and difference frequencies. *Hint: It is a mixer circuit, so there are two input frequencies, and the result of the mixing is a sum and a difference.*

When an excessive amount of signal energy reaches a mixer circuit, spurious mixer products are generated. *Hint: Overloading a circuit produces spurious products.*

PRACTICAL CIRCUITS

Rectification, detection, and demodulation are used interchangeably. The three terms have the same result when referring to receiver functions.

A diode envelope detector functions by rectification and filtering of RF signals. *Hint: A detector detects RF.*

The detector used for demodulating SSB signals is called a product detector.

E7F DSP and Software-Defined Radio

Direct digital conversion in a software-defined radio refers to incoming RF being digitized by an analog-to-digital converter without being mixed with a local oscillator circuit. *Hint: Digital conversion digitizes.* It is direct because the RF is not first converted to an IF with a local oscillator but you don't need to know that to answer the question.

The digital signal processing audio filter used to remove unwanted noise from a received SSB signal is an adaptive filter. *Hint: Digital filters use computing power to look for patterns and adapt to eliminate the noise.*

The digital processing filter used to generate an SSB signal is a Hilbert-transform filter. *Hint: Digital processing is transforming digits to audio. You need a transform filter.*

A common method of generating an SSB signal using digital signal processing is signals are combined in a quadrature phase relationship.

An analog signal must be sampled by an analog-to-digital converter at least twice the rate of the highest frequency component of the signal. To convert from analog to digital requires double the sampling.

Extra Class – The Easy Way

PRACTICAL CIRCUITS

The minimum number of bits required for an analog-to-digital converter to sample with a range of 1 volt at a resolution of 1 millivolt is 10 bits. Solve: The resolution is 1/1000 of the range. It takes 10 bits to count to a thousand. $2^{10} = 1024$.

The function of a Fast Fourier Transform is to convert digital signals from the time domain to the frequency domain. Cheat: If it is fast, it must have something to do with time.

The function of decimation is reducing the effective sample rate by removing samples. When the Romans conquered, they would often kill every tenth enemy captured, a process called decimation. Decimation reduced the enemy's effectiveness by removing a sample. This has nothing to do with radio, but it might help you remember.

An anti-aliasing digital filter is required in a digital decimator to remove high-frequency signal components which would otherwise be reproduced as lower frequency components. Hint: Those lower-frequency components would be aliases of the signal.

In a Direct Digital Conversion SDR receiver using analog-to-digital conversion, the maximum bandwidth is determined by the sample rate. SDR means "Software Defined Radio." Hint: A higher sample rate allows for more conversion, which is more bandwidth.

The minimum detectable signal level for an SDR in the absence of atmospheric or thermal noise is set by the reference voltage level and sample width. Cheat: Can't we say the signal level is set by voltage?

PRACTICAL CIRCUITS

The advantage of a Finite Impulse Response (FIR) filter vs. an Infinite Impulse Response (IIR) filter would be FIR filters delay all frequency components of the signal by the same amount. *Hint: The delay is "finite," a set amount that stays the same.*

The function of taps in a digital processing filter is to provide incremental signal delays for filter algorithms. *Hint: think of "taps" as time-outs, delays that allow the filter to catch up.*

To create a sharper filter response in a digital processing filter, use more taps. *Hint: More time to process.*

E7G– Active Filters and Op-amp Circuits

The typical output impedance of an integrated circuit op-amp is very low. *Hint: Low impedance means lots of current, which means lots of power, and that is what you want in an amplifier.*

The typical input impedance of an integrated circuit op-amp is very high. *Hint: High input impedance means it does not load down the circuit and only draws a little power on the input. Op-amps are high impedance input and low impedance output.*

An operational amplifier is a high-gain, direct-coupled differential amplifier with a very high input impedance and a very low output impedance.

The "op-amp offset voltage" is the differential line voltage needed to bring the open loop output voltage to zero. *Hint: You adjust the offset to bring the device to zero.*

PRACTICAL CIRCUITS

The effect of ringing in a filter is undesired oscillations added to the desired signal. *Hint: The question is about the effect, not the cause. The wrong answer mentions a cause. The effect is undesired oscillations.*

To prevent ringing and audio instability in an op-amp audio filter circuit, restrict both gain and Q. *Hint: Too too much gain will cause instability, and too narrow a filter (high-Q) will ring.*

The gain-bandwidth of an operational amplifier is the frequency at which the open-loop gain of the amplifier equals one.

The gain of an ideal operational amplifier does not vary with frequency. *Hint: The question is about an ideal amplifier, and such an amplifier should not vary with frequency.*

In Figure E7-3, if R1 is 1,000 ohms, RF is 10,000 ohms and 0.23 volts DC is applied to the input, the output voltage would be -2.3 volts. *Hint: The ratio of the resistors RF/R1 is 10 so the gain is 10 times. The output voltage of an op-amp is inverted. Hence the answer is -10 x .23 = -2.3 volts.*

The absolute voltage gain expected when R1 is 1,800 ohms and RF is 68 kilohms is 68,000 / 1,800 = 38. *Hint: RF/R1 = 68,000 / 1,800 = 38. Here the question is about voltage gain, not voltage. There is no inverting, and luckily, there are no answers with negative numbers to distract you.*

The absolute voltage gain expected from the circuit in E7-3 when R1 is 3300 ohms and RF is 47 kilohms ohms is 14. *Solve: It is the ratio of the two resistors. RF/R1 = 47,000/3,300 = about 14.*

PRACTICAL CIRCUITS

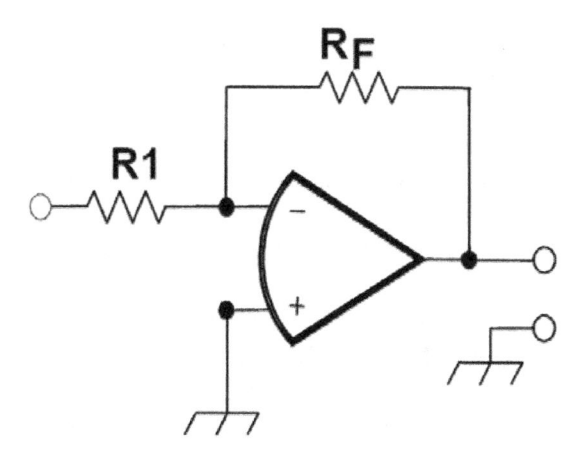

Figure E7-3

Figure E7-3 is an op-amp.

E7H Oscillators and Signal Sources

Three oscillator circuits used in Amateur Radio are Colpitts, Hartley, and Pierce. *Hint: Colpitts is only in the correct answer. Pick the colPitts. One or more of these three circuits are in all the oscillator answers.*

Positive feedback in a Hartley oscillator is supplied through a tapped coil. *Cheat: HarTley and Tapped coil.*

Positive feedback in a Colpitts oscillator is supplied through a capacitive divider. *Cheat: Colpitts and Capacitive.*

Positive feedback in a Pierce oscillator is supplied through a quartz crystal. *Cheat: Mind your Ps and Qs.*

The oscillator circuits used in VFOs are Colpitts and Hartley. *Hint: Pierce uses a crystal, so it*

Extra Class – The Easy Way

PRACTICAL CIRCUITS

*wouldn't do for a VFO. Cheat: VFOs **CH**ange frequency.*

A microphonic is a change in oscillator frequency due to mechanical vibration.

You can reduce an oscillator's microphonic responses by mechanically isolating the oscillator from its enclosure. *Hint: Keep it from mechanical vibration by mechanically isolating it.*

Components to reduce thermal drift in a crystal oscillator are NP0 capacitors. NP0 capacitors are made from ceramic and are very stable, so they don't change value when heated.

The frequency synthesizer that uses a phase accumulator, lookup table, digital to analog converter and a low-pass anti-alias filter is a direct digital synthesizer. *Hint: It has a digital to analog converter, so it must be a digital synthesizer.*

The information contained in the lookup table of a direct digital synthesizer (DDS) is the amplitude values that represent the desired waveform. *Hint: A table would contain "values." Look for the answer with "values."*

The major spectral impurity components of direct digital synthesizers are spurious signals at discrete frequencies. *Hint: Spectral impurity is caused by spurious signals.*

To insure that a crystal oscillator provides the frequency specified by the manufacturer, provide the crystal with a specified parallel capacitance. A crystal is designed for a particular frequency, but it can be "pulled" by stray capacitance. The manufacturer calibrates for a specified parallel capacitance.

PRACTICAL CIRCUITS

A technique for providing highly accurate and stable oscillators for microwave transmission and reception is:
- **Use a GPS signal reference.**
- **Use a rubidium stabilized reference oscillator.**
- **Use a temperature-controlled high Q dielectric resonator.**
- **All these choices are correct.**

Hint: GPS and rubidium are easy to remember, and when you have confidence in 2 out of 3, you can be safe selecting "all of the above."

A phase-locked loop circuit is an electronic servo loop consisting of a phase detector, a low-pass filter, a voltage-controlled oscillator, and a stable reference oscillator. *Hint: If it is "locked," it must have a reference oscillator to lock with. Look for the answer with "reference oscillator" and ditch the rest of this bloated answer.*

The functions that can be performed by a phase-locked loop are frequency synthesis and FM demodulation. *Hint: A loop implies an oscillator, and only one answer has frequency synthesis.*

E8 – SIGNALS AND EMISSIONS

E8A Waveforms

The process that shows a square wave is made up of a sine wave plus all its odd harmonics is Fourier analysis. *Fourier analysis runs the signal through mathematical filters to extract its parts. A square wave contains the odd harmonics.*

A type of analog-to-digital conversion is successive approximation. *Hint: The converter goes through successive stages to calculate the end result.*

Fourier analysis shows a wave made up of sine waves of a given fundamental plus all its harmonics is a sawtooth wave. *Hint: A sawtooth wave includes all harmonics.*

"Dither" in an analog-to-digital converter is a small amount of noise added to the input signal to allow a more precise representation of the true signal over time. *Adding the noise prevents the converter from accumulating errors over time.*

The most accurate way to measure the RMS voltage of a complex waveform is with a true-RMS calculating meter. *Hint: Use a true meter to get the most accurate reading.*

The approximate ratio of PEP-to-average power in a typical single-sideband phone signal is 2.5 to 1. *Hint: Peak is higher than average but not crazy higher.*

The factor that determines the PEP-to-average ratio of a single-sideband signal is the speech

SIGNALS AND EMISSIONS

characteristics. *Hint: Continuous and loud will put out more PEP power.*

A direct or flash conversion analog-to-digital converter would be useful for a software-defined radio because very high speed allows digitizing high frequencies. *Hint: "direct or flash" implies high speed.*

An analog-to-digital converter with an 8 bit resolution can encode 256 levels. *Solve $2^8 = 256$.*

The purpose of a low pass filter used in a digital-to-analog converter is to remove harmonics from the output caused by discrete analog levels generated. *Hint: A low pass filter removes harmonics. You don't care how they were caused.*

A measure of the quality of an analog-to-digital converter is total harmonic distortion. *Hint: Low distortion would be a good thing.*

E8B Modulation and Demodulation

The modulation index of an FM signal is the ratio of frequency deviation to modulating signal frequency. *Hint: A "ratio" is sometimes called an "index." The correct answer compares two frequencies.*

The modulation index of an FM-phone signal having a maximum frequency deviation of 3000 Hz on either side of the carrier frequency when the modulating frequency is 1000 Hz is 3. *Solve: The ratio of deviation (3000) and modulating frequency (1000) is 3.*

The modulation index of an FM-phone signal having a maximum carrier deviation of plus or

SIGNALS AND EMISSIONS

minus 6 kHz when the modulated with a 2 kHz modulating frequency is 3. *Solve: 6/2 = 3.*

Deviation ratio is the ratio of the maximum carrier frequency deviation (swing) **to the highest audio modulating frequency.**

The deviation ratio of an FM-phone signal having a maximum frequency swing of plus-or-minus 5 kHz when the maximum modulating frequency is 3 kHz. is 5000/3000 = 1.67

The deviation ratio of an FM-phone signal having a maximum frequency swing of plus or minus 7.5 kHz when the maximum modulation frequency is 3.5 kHz is 7.5/3.5 = 2.14. *Hint: Calculate deviation ratio in the same manner as the modulation index: divide the larger number by the smaller.*

The modulation index of a phase-modulated emission does not depend on the carrier frequency. *Hint: If it is phase modulated, the carrier frequency doesn't change*

Orthogonal Frequency Division Multiplexing is a technique used for high-speed digital modes. *Hint: "Multiplexing" implies high speed.*

Orthogonal Frequency Division Multiplexing is a digital modulation technique using subcarriers at frequencies chosen to avoid intersymbol interference. *Hint: "multiplexing" means it uses additional subcarriers. Choose the answer with "subcarriers."*

Frequency division multiplexing is two or more information streams merged into a baseband, which then modulates the transmitter. *Hint: Multiplexing is two or more. The word to look for is "baseband."*

SIGNALS AND EMISSIONS

Digital time division multiplexing is two or more signals arranged to share discrete time slots of a data transmission. *Hint: Multiplexing "shares." Look for "time division" and "time slots."*

E8C Digital Signals

Forward Error Correction is implemented by transmitting extra data that may be used to detect and correct transmission errors. *Hint: The purpose of correction is to detect and correct errors.*

The symbol rate in a digital transmission is the rate at which the waveform changes to convey information. *Hint: Look for "rate" in the question and answer.*

Phase-shifting of a PSK signal should be done at the zero crossing of the RF signal because this minimizes bandwidth. *Hint: "Minimize bandwidth" is always good.*

The technique used to minimize bandwidth requirements of a PSK31 signal is use of sinusoidal data pulses. *Hint: Sinusoidal means a smooth and repetitive oscillation. "Smooth" sounds like it would minimize bandwidth. Look for the word "sinusoidal."*

The approximate bandwidth of a 13-WPM international Morse code transmission is 52 Hz.

The necessary bandwidth of a 170-hertz shift 300 baud ASCII transmission is .5k Hz.

The necessary bandwidth of a 4800-Hz frequency shift 9600 baud ASCII FM transmission is 15.36 kHz.
Hint: The answer is always about 4 times the shift or WPM: 13 x 4 = 52; 170 x 4 = about 500;

Extra Class – The Easy Way

SIGNALS AND EMISSIONS

4800 x 4 = about 15.36 kHz.

The way ARQ accomplishes error correction is if errors are detected, retransmission is requested. Hint: ARQ stands for Automatic Repeat reQuest.

The digital code which allows only one bit to change between sequential code values is Gray code. Hint: If you see "Gray code" in an answer, it is correct.

Data rate may be increased without increasing bandwidth by using a more efficient digital code. Hint: More efficient increases the rate.

The relationship between symbol rate and baud is that they are the same. Hint: The two words mean the same thing.

The factors that affect the bandwidth of a transmitted CW signal are keying speed and shape factor (rise and fall time). Hint: Faster digital modes are wider, so is faster CW.

E8D Keying Defects and Overmodulation

Spread spectrum signals are resistant to interference because signals not using the spread spectrum algorithm are suppressed in the receiver. Hint: If a signal doesn't behave like it is expected to, the receiver won't follow it.

The spread spectrum communications technique that uses a high-speed binary bit stream to shift the phase of an RF carrier is called direct sequence. Instead of changing frequency, a binary bit stream of extra data is injected directly into the signal. Caution: The wrong answer is "binary phase-shift keying."

SIGNALS AND EMISSIONS

How the spread spectrum technique of frequency hopping works is the frequency of the transmitted signal is changed very rapidly according to a pseudorandom sequence also used by the receiving station. *Hint: SS hops around with the receiver following the transmitter.*

The primary effect of extremely short rise and fall times on a CW signal is the generation of key clicks.
The most common method of reducing key clicks is to increase the keying waveform rise and fall times.

The advantage of using a parity bit with an ASCII stream is some types of errors can be detected. The parity bit tells the receiver if the ASCII stream had an even or odd number of data bits. It is the simplest form of error detection.

A common cause of overmodulating AFSK signals is excessive transmit audio levels. *Hint: AFSK is audio frequency-shift keying. If you overdrive the audio, you will overmodulate the signal.*

The parameter that evaluates distortion of an AFSK signal caused by excessive input audio levels is Intermodulation Distortion (IMD). *Hint: "Distortion" is in both the question and answer.*

An acceptable maximum IMD level for an idling PSK signal is -30dB. Over-driven audio is very obvious on waterfall displays. You will see a wide display and often, parallel lines.

Some of the differences between Baudot digital code and ASCII are Baudot uses 5 data bits per character, and ASCII uses 7 or 8. *Hint: The answer is longer, but all you need to recognize is that Baudot uses 5 data bits.*

SIGNALS AND EMISSIONS

An advantage to using ASCII for data transmissions is it is possible to transmit both lower and upper case. Baudot is all caps with no lower case letters.

E9 – ANTENNAS AND TRANSMISSION LINES

E9A Basic antenna parameters

An isotropic antenna is a theoretical, omnidirectional antenna used as a reference for antenna gain. *Hint: The word "isotropic" means "equal way." An isotropic antenna is a theoretical single point that radiates equally in all directions.*

The antenna that has no gain in any direction is an isotropic antenna. *Equal in all directions.*

The effective radiated power relative to a dipole of a station with 150 watts transmitter power, 2 dB feed line loss, 2.2dB duplexer loss and 7 dBd of antenna gain is 286 watts. *Solve: First the dBs: 7 dB gain - 2 - 2.2 = 2.8 dB. 3 dB of gain is double the power or 300 watts. The answer is the one that is a little less than 300 = 286 watts.*

The effective radiated power relative to a dipole of a repeater station with 200 watts transmitter output, 4 dB feed line loss, 3.2 dB duplexer loss, .8 dB circulator loss and 10 dBd of antenna gain is 317 watts. *Solve: 10 – 4 – 3.2 - .8 = 2 db. 2 dB is less than double the power and the power ratio of 2 dB is not a nice even whole number so the answer is the 317.*

The effective radiated power of a repeater station with 200 watts transmitter power, 2 dB feed line loss, 2.8 dB duplexer loss, 1.2 dB circulator loss and a 7 dBi antenna gain is 252 watts. *Solve: The total gain is 7 – 2 – 2.8 – 1.2 = 1 dB. The answer is the one just over 200 watts.*

ANTENNAS AND TRANSMISSION LINES

The radiation resistance of an antenna is the value of resistance that would dissipate the same amount of power as that radiated from an antenna. *Hint: "Radiation" and "radiated" in the question and answer.*

The factor that may affect the feed point impedance of an antenna is the antenna height. *Hint: The ground influences the impedance because the antenna capacitively couples to the ground. The higher the antenna, the less the effect.*

The total resistance in an antenna system is the radiation resistance plus loss resistance. *Hint: Loss resistance is resistance in the wire itself. Total resistance would be the antenna plus the wire. Recognize "loss resistance" in the correct answer.*

Antenna bandwidth means the frequency range over which an antenna satisfies a performance requirement. *Hint: It tells you how wide a range the antenna can cover.*

Antenna efficiency is calculated by radiation resistance / total resistance. *Hint: The radiation resistance is what is radiating, the total resistance is everything, including the loss resistance. The more radiating, the better.*

A way to improve the efficiency of a ground-mounted quarter-wave vertical antenna is to install a radial system. *Hint: The radials provide the other half of the dipole and reduce ground losses.*

The factor that determines ground losses for a ground-mounted vertical antenna operating in the 3 MHz to 30 MHz range is soil conductivity. *Hint: Soil is not very conductive and therefore increases the losses.*

ANTENNAS AND TRANSMISSION LINES

Gain in an antenna is often measured in reference to a dipole (dBd) or isotropic (dBi). A gain figure alone, without telling you what it is in reference to, is meaningless. Note it does not depend on transmitter power. The transmitter is producing the same amount of power, the antenna is concentrating that power.

A half-wavelength dipole has about 2.15 dB gain over an isotropic antenna. The following question asks you to compare the two. Subtract 2.15 from the gain over the isotropic antenna to get the dipole gain.

If an antenna has 6 dB gain over an isotropic antenna, it is 3.85 dB better than a 1/2 wavelength dipole *Solve: 6 - 2.15 = 3.85.*

The term that describes station output, taking into account all gains and losses is effective radiated power. You determine the effective radiated power of a system by adding and subtracting the dB loss and gain of each component.

E9B Antenna Patterns

To determine the approximate beam-width in a given plane of a directional antenna, note the two points where the signal strength of the antenna is 3 dB less than maximum and look at the angular difference. That is what you will do in the next question.

The beam-width in the antenna radiation pattern shown in Figure E9-1 above is 50 degrees. *Solve: The pattern hits the -3 dB ring at a little more than 30 degrees and less than -30 degrees so the width is a little less than the total of 60 degrees. 50 degrees is the closest answer.*

ANTENNAS AND TRANSMISSION LINES

Figure E9-1

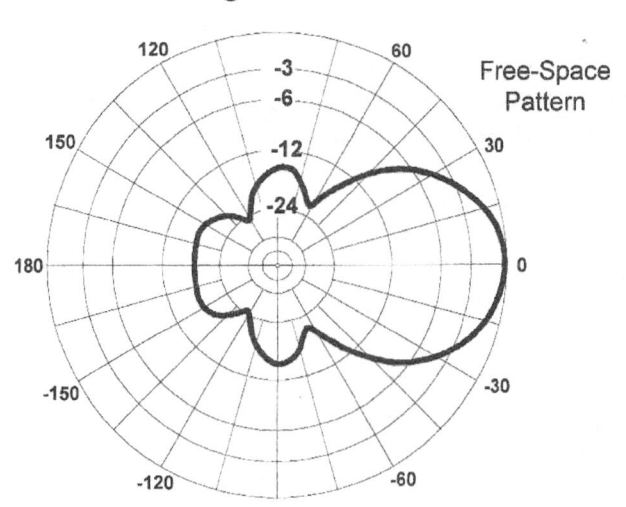

Free-Space Pattern

The front-to-back ratio is 18 dB. *Solve:* The front is at 0 dB, the back is at -18 dB (not well marked but it is halfway between the -12 and -24 circles)

The front-to-side ratio is -14 dB. *Solve:* The front is 0 dB, and the side is -14 dB (again, not well marked but if you look carefully, you see it is closer to the -12 circle).

The next questions refer to Figure 9-2 on the following page.

In Figure E9-2 the antenna pattern is an elevation pattern. It shows the takeoff angles.

The front-to-back ratio shown in Figure E9-2 is 28 dB. It isn't well marked but you can see the back is just a little over the -30 dB arc.

ANTENNAS AND TRANSMISSION LINES

Figure E9-2

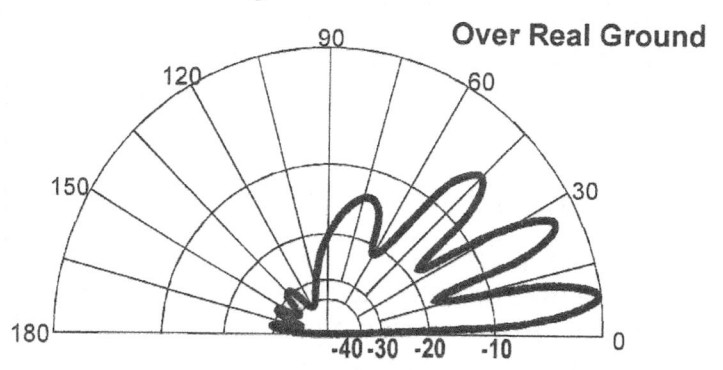

Over Real Ground

The elevation level of the peak response is 7.5 degrees. *The strongest lobe is the bottom at 7.5 degrees.*

The total amount of radiation emitted by a directional gain antenna compared with the total amount of radiation from an isotropic antenna assuming each is driven by the same amount of power, is the same. *Hint: The total amount of radiation is the same, the directional antenna concentrates it.*

The far field of an antenna is the region where the shape of the antenna pattern is independent of distance. *Hint: You are far enough away that the pattern is no longer affected by ground or objects.*

Antenna modeling programs use a computer program technique called Method of Moments.

The principle of a Method of Moments analysis is a wire is modeled as a series of segments, each having a uniform value of current. *Hint: Current is what causes the antenna to radiate.*

Extra Class – The Easy Way

ANTENNAS AND TRANSMISSION LINES

The disadvantage of decreasing the number of wire segments in an antenna model below 10 segments per half-wavelength is the computed feed point impedance may be incorrect. *Hint: Fewer data points mean less precision in the answer. You don't need to know how many.*

E9C Wire and Phased Array Antennas

The radiation pattern of two 1/4-wavelength vertical antennas spaced 1/2-wavelength apart and fed 180 degrees out of phase is a figure-8 oriented along the axis of the array.

The radiation pattern of two 1/4-wavelength vertical antennas spaced 1/4- wavelength apart and fed 90 degrees out of phase is a cardioid.

The radiation pattern of two 1/4-wavelngth vertical antennas spaced 1/2-wavelength apart and fed in phase is a figure-8 broadside to the axis of the array.

Memory point:
180 out of phase, oriented along.
90 out of phase, cardioid.
In phase, broadside.

As the wire length is increased for an unterminated long wire antenna, the radiation pattern changes so the lobes align more in the direction of the wire. *Hint: A long wire antenna radiates in the direction of the wire.*

An OCFD antenna is fed approximately 1/3 of the way from the end and with a 4:1 balun to provide multiband operation. *Hint: OCFD stands for "off-center-fed-dipole." It is fed 1/3 of the way from the end. That is all you need to recognize to answer the question.*

ANTENNAS AND TRANSMISSION LINES

A rhombic antenna is two long wires pointed in the same direction in the shape of a rhombus. **The effect of a terminating resistor on a rhombic antenna is to change the radiation pattern from bidirectional to unidirectional.**

A folded dipole antenna is a half-wave dipole with an additional parallel wire connecting its two ends. *Hint: It folds back on itself.*

A two-wire folded dipole antenna has a feed point impedance at the center of 300 ohms.

A G5RV antenna is a multi-band dipole fed with coax and a balun through a selected length of open wire transmission line. *Hint: the G5RV is a multi-band dipole. That is all you need to know.*

A Zepp antenna is an end-fed dipole. *Hint: Zepp antennas got their name because they were a wire trailing behind a Zeppelin and, fed at the end.*

An Extended Double Zepp antenna is a center fed 1.25 wavelength antenna. *Hint: It is extended, so it is extra long. This Zepp is not end fed.*

The effect on the far-field elevation pattern of a vertically polarized antenna mounted over seawater versus soil is that the low-angle radiation increases. *Hint: Far-off DX signals arrive at a low angle. Verticals over saltwater are great low-angle DX antennas.*

The radiation pattern of a 3-element beam antenna varies with height in that the takeoff angle decreases with increasing height. *Hint: The signal starts higher and therefore travels further before it hits the ground. It bounces off the ground at a lower angle.*

ANTENNAS AND TRANSMISSION LINES

A horizontally polarized antenna mounted on a hill vs one mounted on flat ground will differ in that **the takeoff angle decreases in the downhill direction.** *Hint: The signal bounces off the hill further away and at a lower elevation.*

E9D Directional and Short Antennas

The gain of an ideal parabolic dish when the operating frequency is doubled is increased by **6 dB.** *Solve: Gain increases with the square of the ratio of the aperture width to wavelength. Doubling the frequency increases the gain by $2^2 = 4$ times or 6 dB. Careful! The answer is not 4 dB.*

Linearly polarized Yagi antennas can be made to produce circular polarization by arranging the 2 Yagis perpendicular to each other with the driven elements at the same point fed 90 degrees out of phase. *Hint: Mount the two yagis perpendicular to each other.*

Coils are often used to lengthen a short antenna electrically. **To minimize losses on a shortened vertical antenna, a high Q loading coil should be placed near the center of the vertical radiator.**

An HF mobile antenna should have a high ratio of reactance to resistance to minimize losses. *Hint: Higher resistance would mean higher losses.*

If a Yagi antenna is designed solely for forward gain, the front-to-back ratio decreases. *Yagi design is a trade-off.*

When one or more loading coils are used to resonate an electrically short antenna, the SWR bandwidth is decreased. *Shortened antennas can have very narrow bandwidth.*

ANTENNAS AND TRANSMISSION LINES

An advantage of using top loading in a shortened vertical antenna is improved radiation efficiency *Hint: The maximum current is at the base. Let the antenna radiate before it suffers losses in a coil or loading.*

As the Q of an antenna increases, the SWR bandwidth decreases. *Hint: Higher Q means sharper tuning and bandwidth decreases.*

The function of a loading coil used as part of an HF mobile antenna is to cancel capacitive reactance. *Hint: Inductance cancels capacitance.*

The effect on the feed-point impedance at the base of a fixed length HF mobile antenna, when operated below its resonant frequency, is the radiation resistance decreases and the capacitive reactance increases. *Hint: The antenna becomes less efficient – the radiation resistance decreases. The antenna becomes too short – it has capacitive reactance.*

Conductors best for minimizing losses in a station's RF ground system would be wide, flat copper strap. *Hint: Skin effect means the RF flows on the surface, and flat strap has more surface than round wire.*

The best RF ground for your station would be an electrically short connection to 3 or 4 interconnected ground rods driven into the Earth. *Hint: Ground to the ground (Earth).*

E9E Matching

The system that matches a higher impedance feed line to a lower impedance antenna by connecting the line to the driven element in two places spaced a fraction of a wavelength each

Extra Class – The Easy Way   Page 95

ANTENNAS AND TRANSMISSION LINES

side of the element center is called a delta matching system. *Hint: Feeding away from the center on both sides would look like a triangle (delta).*

The antenna matching system that matches an unbalanced feed line to an antenna by feeding the driven element both at the center of the element and at a fraction to one side of center is called a gamma match. *Hint: Fed in the center and another place is gamma.*

An effective method of shunt-feeding a grounded tower at its base is a gamma match. *Hint: Another use for the gamma match.*

The matching system that uses a section of transmission line connected in parallel with the feed line at or near the feed point is called stub match. *Hint: The parallel feed line is a stub.*

The purpose of a series capacitor in a gamma-type matching network is to cancel the inductive reactance of the matching network. *Hint: Capacitors cancel inductance.*

Another matching scheme is called hairpin matching. **The driven element in a 3-element Yagi tuned to use a hairpin matching system, should be capacitive.** A little short. *Hint: Hairpins are short.*

The feed line impedance suitable for constructing a quarter-wave Q-section for matching a 100-ohm loop to a 50-ohm feed line would be 75 ohms. One-quarter wavelength of coax can act as an impedance transformer. Multiply the two impedances and take the square root of the answer. *Hint: Or remember this common trick.*

An effective choice to match an antenna with a 100-ohm feed point impedance to a 50-ohm

ANTENNAS AND TRANSMISSION LINES

coaxial cable feed would be to **insert a ¼-wavelength piece of 75-ohm coaxial cable transmission line in series between the antenna terminals and the 50-ohm feed cable.** *Hint: Same question as above if you remember ¼-wave, 75 ohm.*

The term to describe the interaction at the load end of a mismatched transmission line is **reflection coefficient.** *Hint: A mismatched line will reflect power back.*

A use for a Wilkinson divider is to **divide power equally between two 50-ohm loads while maintaining 50-ohm input impedance.** *Hint: It is a divider, so it divides power. The other answer about dividing frequency doesn't make any sense.*

The primary purpose of phasing lines, when used with an antenna having multiple driven elements, is **it ensures that each driven element operates in concert with the others to create the desired antenna pattern.** *Hint: You want to create the desired antenna pattern.*

E9F Transmission Lines

The velocity factor of a transmission line is **the velocity of the wave in the transmission line divided by the velocity of light in a vacuum.** *Hint: It is the ratio of speed in the line compared to the speed of light in a vacuum. The velocity factor is never more than 1.*

The velocity factor of a transmission line is determined by **the dielectric material used in the line.** The type of insulation.

The physical length of a coaxial cable is shorter than its electrical length because electrical

ANTENNAS AND TRANSMISSION LINES

signals move more slowly in a coaxial cable than in air.

If you are cutting a line for a certain frequency, you have to take the velocity factor into account. The velocity factor for polyethylene dielectric coax is about .67. So cut it to 67% of the calculated length, and the wave will arrive at the right time.

The approximate physical length of a solid polyethylene dielectric coaxial cable that is electrically one-quarter wavelength at 14.1 MHz is 3.5 meters. *Solve in your head: The velocity factor of solid dielectric coax is about .67. The frequency is 20 meters, so that one-quarter wavelength would be 5 meters. 5 X .67 = 3.35 and 3.5 is the closest answer.*
Solve with a calculator. To find the wavelength, divide 300 by the frequency. The wavelength is 300/14.1 or 21.3 meters. One-quarter would be 5.32 meters times .67 = 3.56 meters.

The approximate physical length of an air-insulated parallel transmission line that is electrically one-half wavelength long at 14.1 MHz would be 10 meters.
Hint: Air-insulated line has no dielectric, so the velocity factor is very close to 1 (.95 actually).
Solve: The frequency is 20 meters, and one-half wavelength would be 10 meters.
Solve with a calculator: We know the frequency is 21.3 meters, and half of that is 10.65. The velocity factor of the line is .95 so 10.65 x .95 = 10.1 meters

Ladder line compares to small diameter coaxial cable such as RG-58 at 50 MHz in that the ladder line has lower loss. *Hint: Ladder line has much lower losses than coax at all frequencies.*

ANTENNAS AND TRANSMISSION LINES

The significant difference between foam dielectric coaxial cable and solid dielectric cable is:
- Foam has lower safe operating voltage limits.
- Foam has lower loss.
- Foam has higher velocity factor.
- **All of the choices are correct.**

Foam dielectric coax has a velocity factor of about .85. **The approximate physical length of a foam polyethylene dielectric coaxial cable that is electrically one-quarter wavelength at 7.2 MHz is 8.3 meters.** *Solve: The frequency is 40 meters so one-quarter is 10 meters. 10 x .85 = about 8.3 meters.*
Solve with a calculator: The wavelength is 300/7.2 = 41.66 meters. One-quarter is 10.42. 10.42 x .85 = 8.85 meters. 8.3 is the closest answer.

Memory point:
Solid dielectric velocity factor .67
Foam dielectric .85

Transmission line can act as a transformer. A half-wave line mirrors the end impedance. **The impedance of a 1/2 wavelength transmission line when the line is shorted at the far end is a very low impedance.** *Hint: Same as the end.*

A quarter-wave transmission line inverts the impedance. **The impedance of a 1/4-wavelength line when the line is open at the far end is very low impedance.** *Hint: Opposite.*
The impedance of a 1/4- wavelength transmission line when the line is shorted at the far end is very high impedance. *Hint: Opposite.*

An open or shorted 1/8-wavelength transmission line results in a capacitive or inductive reactance. **The**

ANTENNAS AND TRANSMISSION LINES

impedance of a 1/8-wavelength transmission line when the line is open at the far end is **capacitive reactance**. *Cheat: Think of the center and shield as parallel plates of a capacitor.*

The impedance of a 1/8 wave transmission line when the line is shorted at the far end is an **inductive reactance**. *Cheat: The opposite of above.*

E9G Smith Chart

Using a Smith Chart, you can calculate **impedance along transmission lines**. *Hint: The impedance seen at the feed point changes along the length of a transmission line, as shown above. The Smith Chart is a way of computing without using incredibly complex mathematical equations.*

The coordinate system on a Smith Chart is **resistance circles and reactance arcs**. *Hint: Impedance is resistance and reactance. The chart is a circle. Resistance circles.*

The two families of circles that make up a Smith Chart are **resistance and reactance**.

A Smith Chart determines **impedance and SWR** values in transmission lines. *Hint: If impedance changes, so does SWR.*

A common use of the Smith Chart is to determine the **length and position of an impedance matching stub**.

The name of the large outer circle on which the reactance arcs terminate is the **reactance axis**. *Hint: It is showing the value of the reactance on that arc. Reactance arcs relate to the reactance axis.*

ANTENNAS AND TRANSMISSION LINES

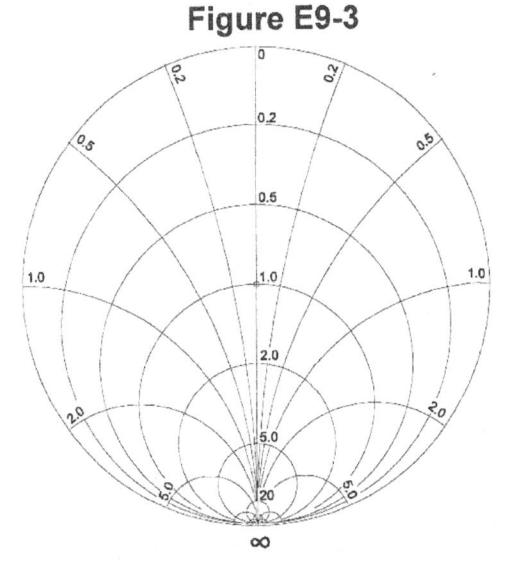

Figure E9-3

The chart shown in E9-3 is a Smith Chart.

The arcs on a Smith Chart represent points with a constant reactance. *Hint: Arcs are reactance.*

The only straight line shown is called the resistance axis. *Hint: Resistance and reactance. Reactance is arcs. Resistance is the other axis.*

The process of normalization on a Smith Chart is reassigning impedance values with regard to the prime center. *Hint: You make it normal by reassigning to the prime center.*

The third family of circles often added to a Smith Chart during the process of solving problems are standing wave ratio circles.

The wavelength scales on a Smith Chart are calibrated in fractions of transmission line electrical wavelength. *Hint: You are calculating a transmission line depending on its length.*

Extra Class – The Easy Way

ANTENNAS AND TRANSMISSION LINES

E9H Receiving Antennas

When constructing a Beverage antenna, the factor to include in the design to achieve good performance is it should be more than one wavelength long. *Hint: Beverages are long and low. They minimize noise on receive.*

Generally, on a low band (160 meters and 80 meters) receiving antenna, the atmospheric noise is so high that gain over a dipole is not important. *Hint: In fact, gain might make the noise worse.*

Receiving Directivity Factor (RDF) is forward gain compared to average gain over the entire hemisphere. *Hint: It is not front-to-back but front to everywhere else.*

The advantage of placing a grounded electrostatic shield around a small loop direction-finding antenna is it eliminates unbalanced capacitive coupling to the surroundings, improving the nulls. *Hint: When using a direction-finding antenna, you turn it to find the null in the signal because a null is easier to discern than a peak. "Improving the nulls" is an advantage and all you need to recognize to answer.*

A small loop antenna will have a sharp null. **The main drawback of a small wire-loop antenna for direction finding is that it is bidirectional.** *Hint: It is a drawback to hear nulls in both front and back.*

The triangulation method of direction finding is antenna headings from several different receiving locations are used to locate the signal source. *Hint: Draw the headings from several receiving locations on a map and where the lines cross is the location of the signal source.*

ANTENNAS AND TRANSMISSION LINES

RF attenuation is used when direction-finding to prevent receiver overload which reduces pattern nulls. *Hint: You put in enough attenuation to reduce the signal to almost nothing. Then when you point at a null, the signal will disappear completely, making the null more obvious.*

The function of a sense antenna is it modifies the pattern of DF antenna array to provide a null in one direction. *Hint: A null in one direction is preferable to two.*

A Pennant antenna is a small, vertically oriented receiving antenna consisting of a triangular loop terminated in approximately 900 ohms. *Hint: Too much information. Remember it is a receiving antenna, or that it is triangular, or terminated with 900 ohms.*

The output voltage of a multiple-turn receiving loop antenna can be increased by increasing the number of turns and/or the area. *Hint: Bigger is better.*

The feature of a cardioid pattern antenna that makes it useful for direction finding is a very sharp single null. *Hint: A very sharp single null would be the best.*

E0 – SAFETY

E0A Safety

The primary function of an external earth connection or ground rod is lightning protection. *Hint: You want to dissipate the lightning force outside before it gets in the house.*

MPE limits are the Maximum Permissible Exposure to electromagnetic radiation. **When evaluating RF exposure levels from your station to your neighbor's house, you must make sure the signals from your station are less than the uncontrolled Maximum Permitted Exposure limits.** *Hint: Your neighbor can't control what you are doing, so he is entitled to the protection of the lower, uncontrolled exposure limits. You are evaluating exposure levels, not emission levels.*

The range of frequencies over which the FCC human body exposure limits are the most restrictive are 30 – 300 MHz. *Hint: VHF is the most dangerous. Very High Fry-ability.*

When evaluating a site with multiple transmitters operating at the same time, the operators and licensees of transmitters responsible for mitigating over-exposure are each transmitter that produces 5% or more of its MPE limit in areas where the total MPE is exceeded. *Hint: Very confusing answer. Look for the answer with 5%. Practically everyone is responsible for mitigating an unsafe situation.*

A potential hazard of using microwaves in the amateur radio bands is the high gain antennas commonly used can result in high exposure levels. *Hint: Microwaves heat.*

SAFETY

An injury from high-power UHF or microwave transmitters would be localized heating of the body from RF exposure in excess of the MPE limits.

There are separate electric (E) and magnetic (H) field MPE limits because:
- **The body reacts to electromagnetic radiation from both fields.**
- **Ground reflections and scattering make the field impedance vary with location.**
- **E field and H field radiation intensity peaks can occur at different locations.**
- **All these choices are correct.**

SAR measures the rate at which RF energy is absorbed by the body. *Hint: SAR means Specific Absorption Rate or "Soon about to roast."*

Dangerous levels of carbon monoxide from an emergency generator can be detected only by a carbon monoxide detector. *Hint: Carbon monoxide (CO) is colorless and odorless.*

The type of insulating material commonly used as a thermal conductor that is extremely toxic if broken or crushed and the particles are inhaled is beryllium oxide.

The toxic material that may be present in some electronic components such as high voltage capacitor and transformers is Polychlorinated Biphenyls. *Hint: Also called PCBs.*

AMATEUR RADIO EXTRA CLASS QUICK SUMMARY

NOTE: The question pool is divided into Groups within Subelements. There is one question from each of the 50 Groups. If you have trouble with a Group, don't worry. You can only get one question per Group, and you may know the answer to the one.
The Figures (diagrams) are at the back of the book.

E1 — COMMISSION RULES
[6 Exam Questions — 6 Groups]

E1A Operating Standards

E1A01 Which of the following carrier frequencies is illegal for LSB AFSK emissions on the 17 meter band RTTY and data segment of 18.068 to 18.110 MHz?
18.068 MHz

E1A02 When using a transceiver that displays the carrier frequency of phone signals, which of the following displayed frequencies represents the lowest frequency at which a properly adjusted LSB emission will be totally within the band?
3 kHz above the lower band edge

E1A03 What is the maximum legal carrier frequency on the 20 meter band for transmitting USB AFSK digital signals having a 1 kHz bandwidth?
14.149 MHz

E1A04 With your transceiver displaying the carrier frequency of phone signals, you hear a DX station calling CQ on 3.601 MHz LSB. Is it legal to return the call using lower sideband on the same frequency?
No, the sideband will extend beyond the edge of the phone band segment

SUMMARY – COMMISSION RULES

E1A05 What is the maximum power output permitted on the 60 meter band?
100 watts PEP effective radiated power relative to the gain of a half-wave dipole

E1A06 Where must the carrier frequency of a CW signal be set to comply with FCC rules for 60 meter operation?
At the center frequency of the channel

E1A07 What is the maximum power permitted on the 2200 meter band?
1 watt EIRP (Equivalent isotropic radiated power)

E1A08 If a station in a message forwarding system inadvertently forwards a message that is in violation of FCC rules, who is primarily accountable for the rules violation?
The control operator of the originating station

E1A09 What action or actions should you take if your digital message forwarding station inadvertently forwards a communication that violates FCC rules?
Discontinue forwarding the communication as soon as you become aware of it

E1A10 If an amateur station is installed aboard a ship or aircraft, what condition must be met before the station is operated?
Its operation must be approved by the master of the ship or the pilot in command of the aircraft

E1A11 Which of the following describes authorization or licensing required when operating an amateur station aboard a U.S.-registered vessel in international waters?
Any FCC-issued amateur license

E1A12 What special operating frequency restrictions are imposed on slow scan TV transmissions?
They are restricted to phone band segments

SUMMARY – COMMISSION RULES

E1A13 Who must be in physical control of the station apparatus of an amateur station aboard any vessel or craft that is documented or registered in the United States?
Any person holding an FCC issued amateur license or who is authorized for alien reciprocal operation

E1A14 Except in some parts of Alaska, what is the maximum power permitted on the 630 meter band?
5 watts EIRP

E1B Station Restrictions

E1B01 Which of the following constitutes a spurious emission?
An emission outside the signal's necessary bandwidth that can be reduced or eliminated without affecting the information transmitted

E1B02 Which of the following is an acceptable bandwidth for Digital Radio Mondiale (DRM) based voice or SSTV digital transmissions made on the HF amateur bands?
3 kHz

E1B03 Within what distance must an amateur station protect an FCC monitoring facility from harmful interference?
1 mile

E1B04 What must be done before placing an amateur station within an officially designated wilderness area or wildlife preserve or an area listed in the National Register of Historic Places?
An Environmental Assessment must be submitted to the FCC

E1B05 What is the National Radio Quiet Zone?
An area surrounding the National Radio Astronomy Observatory

SUMMARY – COMMISSION RULES

E1B06 Which of the following additional rules apply if you are installing an amateur station antenna at a site at or near a public use airport?
You may have to notify the Federal Aviation Administration and register it with the FCC as required by Part 17 of the FCC rules

E1B07 To what type of regulations does PRB-1 apply?
State and local zoning

E1B08 What limitations may the FCC place on an amateur station if its signal causes interference to domestic broadcast reception, assuming that the receivers involved are of good engineering design?
The amateur station must avoid transmitting during certain hours on frequencies that cause the interference

E1B09 Which amateur stations may be operated under RACES rules?
Any FCC-licensed amateur station certified by the responsible civil defense organization for the area served

E1B10 What frequencies are authorized to an amateur station operating under RACES rules?
All amateur service frequencies authorized to the control operator

E1B11 What does PRB-1 require of regulations affecting amateur radio?
Reasonable accommodations of amateur radio must be made

E1B12 What must the control operator of a repeater operating in the 70 cm band do if a radiolocation system experiences interference from that repeater?
Cease operation or make changes to the repeater to mitigate the interference

SUMMARY – COMMISSION RULES

E1C Control

E1C01 What is the maximum bandwidth for a data emission on 60 meters?
2.8 kHz

E1C02 Which of the following types of communications may be transmitted to amateur stations in foreign countries?
Communications incidental to the purpose of the amateur service and remarks of a personal nature

E1C03 How do the control operator responsibilities of a station under automatic control differ from one under local control?
Under automatic control the control operator is not required to be present at the control point

E1C04 What is meant by IARP?
An international amateur radio permit that allows U.S. amateurs to operate in certain countries of the Americas

E1C05 When may an automatically controlled station originate third party communications?
Never

E1C06 Which of the following is required in order to operate in accordance with CEPT rules in foreign countries where permitted?
You must bring a copy of FCC Public Notice DA 16-1048

E1C07 At what level below a signal's mean power level is its bandwidth determined according to FCC rules?
26 dB

E1C08 What is the maximum permissible duration of a remotely controlled station's transmissions if its control link malfunctions?
3 minutes

SUMMARY – COMMISSION RULES

E1C09 What is the highest modulation index permitted at the highest modulation frequency for angle modulation below 29.0 MHz?
1.0

E1C10 What is the permitted mean power of any spurious emission relative to the mean power of the fundamental emission from a station transmitter or external RF amplifier installed after January 1, 2003 and transmitting on a frequency below 30 MHz?
At least 43 dB below

E1C11 Which of the following operating arrangements allows an FCC-licensed U.S. citizen to operate in many European countries, and alien amateurs from many European countries to operate in the U.S.?
CEPT agreement

E1C12 On what portion of the 630 meter band are phone emissions permitted?
The entire band

E1C13 What notifications must be given before transmitting on the 630 meter or 2200 meter bands?
Operators must inform the Utilities Technologies Council of their call sign and coordinates of the station

E1C14 How long must an operator wait after filing a notification with the Utilities Technologies Council before operating on the 2200 meter or 630 meter band?
Operators may operate after 30 days, providing they have not been told that their station is within 1 km of PLC systems using those frequencies

E1D Satellites

E1D01 What is the definition of telemetry?
One-way transmission of measurements at a distance from the measuring instrument

SUMMARY – COMMISSION RULES

E1D02 Which of the following may transmit special codes intended to obscure the meaning of messages?
Telecommand signals from a space telecommand station

E1D03 What is a space telecommand station?
An amateur station that transmits communications to initiate, modify or terminate functions of a space station

E1D04 Which of the following is required in the identification transmissions from a balloon-borne telemetry station?
Call sign

E1D05 What must be posted at the station location of a station being operated by telecommand on or within 50 km of the earth's surface?
A photocopy of the station license
A label with the name, address, and telephone number of the station licensee
A label with the name, address, and telephone number of the control operator
All these choices are correct

E1D06 What is the maximum permitted transmitter output power when operating a model craft by telecommand?
1 watt

E1D07 Which HF amateur bands have frequencies authorized for space stations?
Only the 40, 20, 17, 15, 12, and 10 meter bands

E1D08 Which VHF amateur bands have frequencies authorized for space stations?
2 meters

E1D09 Which UHF amateur bands have frequencies authorized for space stations?
70 cm and 13 cm

SUMMARY – COMMISSION RULES

E1D10 Which amateur stations are eligible to be telecommand stations of space stations (subject to the privileges of the class of operator license held by the control operator of the station)?
Any amateur station so designated by the space station licensee

E1D11 Which amateur stations are eligible to operate as Earth stations?
Any amateur station, subject to the privileges of the class of operator license held by the control operator.

E1D12 Which of the following amateur stations may transmit one-way communications?
A space station, beacon station, or telecommand station

E1E Examiners

E1E01 For which types of out-of-pocket expenses do the Part 97 rules state that VEs and VECs may be reimbursed?
Preparing, processing, administering, and coordinating an examination for an amateur radio operator license

E1E02 Who does Part 97 task with maintaining the pools of questions for all U.S. amateur license examinations?
The VECs

E1E03 What is a Volunteer Examiner Coordinator?
An organization that has entered into an agreement with the FCC to coordinate, prepare, and administer amateur operator license examinations

E1E04 Which of the following best describes the Volunteer Examiner accreditation process?
The procedure by which a VEC confirms that the VE applicant meets FCC requirements to serve as an examiner

E1E05 What is the minimum passing score on all amateur operator license examinations?
Minimum passing score of 74%

SUMMARY – COMMISSION RULES

E1E06 Who is responsible for the proper conduct and necessary supervision during an amateur operator license examination session?
Each administering VE

E1E07 What should a VE do if a candidate fails to comply with the examiner's instructions during an amateur operator license examination?
Immediately terminate the candidate's examination

E1E08 To which of the following examinees may a VE not administer an examination?
Relatives of the VE as listed in the FCC rules

E1E09 What may be the penalty for a VE who fraudulently administers or certifies an examination?
Revocation of the VE's amateur station license grant and the suspension of the VE's amateur operator license grant

E1E10 What must the administering VEs do after the administration of a successful examination for an amateur operator license?
They must submit the application document to the coordinating VEC according to the coordinating VEC instructions

E1E11 hat must the VE team do if an examinee scores a passing grade on all examination elements needed for an upgrade or new license?
Three VEs must certify that the examinee is qualified for the license grant and that they have complied with the administering VE requirements

E1E12 What must the VE team do with the application form if the examinee does not pass the exam?
Return the application document to the examinee

SUMMARY – COMMISSION RULES

E1F Miscellaneous Rules

E1F01 On what frequencies are spread spectrum transmissions permitted?
Only on amateur frequencies above 222 MHz

E1F02 What privileges are authorized in the U.S. to persons holding an amateur service license granted by the government of Canada?
The operating terms and conditions of the Canadian amateur service license, not to exceed U.S. Amateur Extra Class license privileges

E1F03 Under what circumstances may a dealer sell an external RF power amplifier capable of operation below 144 MHz if it has not been granted FCC certification?
It was purchased in used condition from an amateur operator and is sold to another amateur operator for use at that operator's station

E1F04 Which of the following geographic descriptions approximately describes "Line A"?
A line roughly parallel to and south of the border between the U.S. and Canada

E1F05 Amateur stations may not transmit in which of the following frequency segments if they are located in the contiguous 48 states and north of Line A?
420 MHz - 430 MHz

E1F06 Under what circumstances might the FCC issue a Special Temporary Authority (STA) to an amateur station?
To provide for experimental amateur communications

E1F07 When may an amateur station send a message to a business?
When neither the amateur nor his or her employer has a pecuniary interest in the communications

SUMMARY – COMMISSION RULES

E1F08 Which of the following types of amateur station communications are prohibited?
Communications transmitted for hire or material compensation, except as otherwise provided in the rules

E1F09 Which of the following conditions apply when transmitting spread spectrum emissions?
A station transmitting SS emission must not cause harmful interference to other stations employing other authorized emissions
The transmitting station must be in an area regulated by the FCC or in a country that permits SS emissions
The transmission must not be used to obscure the meaning of any communication
All these choices are correct

E1F10 Who may be the control operator of an auxiliary station?
Only Technician, General, Advanced or Amateur Extra Class operators

E1F11 Which of the following best describes one of the standards that must be met by an external RF power amplifier if it is to qualify for a grant of FCC certification?
It must satisfy the FCC's spurious emission standards when operated at the lesser of 1500 watts or its full output power

E2 - OPERATING PROCEDURES
[5 Exam Questions - 5 Groups]

E2A Amateur Radio in Space

E2A01 What is the direction of an ascending pass for an amateur satellite?
From south to north

E2A02 Which of the following occurs when a satellite is using an inverting linear transponder?
Doppler shift is reduced because the uplink and downlink shifts are in opposite directions
Signal position in the band is reversed
Upper sideband on the uplink becomes lower sideband on the downlink, and vice versa
All these choices are correct

E2A03 How is the signal inverted by an inverting linear transponder?
The signal is passed through a mixer and the difference rather than the sum is transmitted

E2A04 What is meant by the term "mode" as applied to an amateur radio satellite?
The satellite's uplink and downlink frequency bands

E2A05 What do the letters in a satellite's mode designator specify?
The uplink and downlink frequency ranges

E2A06 What are Keplerian elements?
Parameters that define the orbit of a satellite

E2A07 Which of the following types of signals can be relayed through a linear transponder?
FM and CW
SSB and SSTV

Extra Class – The Easy Way   Page 117

SUMMARY – OPERATING PROCEDURES

PSK and packet
All these choices are correct

E2A08 Why should effective radiated power to a satellite that uses a linear transponder be limited?
To avoid reducing the downlink power to all other users

E2A09 What do the terms "L band" and "S band" specify regarding satellite communications?
The 23 centimeter and 13 centimeter bands

E2A10 What type of satellite appears to stay in one position in the sky?
Geostationary

E2A11 What type of antenna can be used to minimize the effects of spin modulation and Faraday rotation?
A circularly polarized antenna

E2A12 What is the purpose of digital store-and-forward functions on an amateur radio satellite?
To store digital messages in the satellite for later download by other stations

E2A13 Which of the following techniques is normally used by low Earth orbiting digital satellites to relay messages around the world?
Store-and-forward

E2B Television

E2B01 How many times per second is a new frame transmitted in a fast-scan (NTSC) television system?
30

E2B02 How many horizontal lines make up a fast-scan (NTSC) television frame?
525

SUMMARY – OPERATING PROCEDURES

E2B03 How is an interlaced scanning pattern generated in a fast-scan (NTSC) television system?
By scanning odd numbered lines in one field and even numbered lines in the next

E2B04 How is color information sent in analog SSTV?
Color lines are sent sequentially

E2B05 Which of the following describes the use of vestigial sideband in analog fast-scan TV transmissions?
Vestigial sideband reduces bandwidth while allowing for simple video detector circuitry

E2B06 What is vestigial sideband modulation?
Amplitude modulation in which one complete sideband and a portion of the other are transmitted

E2B07 What is the name of the signal component that carries color information in NTSC video?
Chroma

E2B08 What technique allows commercial analog TV receivers to be used for fast-scan TV operations on the 70 cm band?
Transmitting on channels shared with cable TV

E2B09 What hardware, other than a receiver with SSB capability and a suitable computer, is needed to decode SSTV using Digital Radio Mondiale (DRM)?
No other hardware is needed

E2B10 What aspect of an analog slow-scan television signal encodes the brightness of the picture?
Tone frequency

E2B11 What is the function of the Vertical Interval Signaling (VIS) code sent as part of an SSTV transmission?
To identify the SSTV mode being used

SUMMARY – OPERATING PROCEDURES

E2B12 What signals SSTV receiving software to begin a new picture line?
Specific tone frequencies

E2C Operating Methods: Contest and DX

E2C01 What indicator is required to be used by U.S.-licensed operators when operating a station via remote control and the remote transmitter is located in the U.S.?
No additional indicator is required

E2C02 Which of the following best describes the term "self-spotting" in connection with HF contest operating?
The often-prohibited practice of posting one's own call sign and frequency on a spotting network

E2C03 From which of the following bands is amateur radio contesting generally excluded?
30 meters

E2C04 Which of the following frequencies are sometimes used for amateur radio mesh networks?
Frequencies shared with various unlicensed wireless data services

E2C05 What is the function of a DX QSL Manager?
To handle the receiving and sending of confirmation cards for a DX station

E2C06 During a VHF/UHF contest, in which band segment would you expect to find the highest level of SSB or CW activity?
In the weak signal segment of the band, with most of the activity near the calling frequency

E2C07 What is the Cabrillo format?
A standard for submission of electronic contest logs

SUMMARY – OPERATING PROCEDURES

E2C08 Which of the following contacts may be confirmed through the U.S. QSL bureau system?
Contacts between a U.S. station and a non-U.S. station

E2C09 What type of equipment is commonly used to implement an amateur radio mesh network?
A wireless router running custom firmware

E2C10 Why might a DX station state that they are listening on another frequency?
**Because the DX station may be transmitting on a frequency that is prohibited to some responding stations
To separate the calling stations from the DX station
To improve operating efficiency by reducing interference
All these choices are correct**

E2C11 How should you generally identify your station when attempting to contact a DX station during a contest or in a pileup?
Send your full call sign once or twice

E2C12 What technique do individual nodes use to form a mesh network?
Discovery and link establishment protocols

E2D Operating methods: VHF and UHF

E2D01 Which of the following digital modes is designed for meteor scatter communications?
MSK144

E2D02 Which of the following is a good technique for making meteor scatter contacts?
**15-second timed transmission sequences with stations alternating based on location
Use of special digital modes
Short transmissions with rapidly repeated call signs and signal reports
All these choices are correct**

Extra Class – The Easy Way   Page 121

SUMMARY – OPERATING PROCEDURES

E2D03 Which of the following digital modes is especially useful for EME communications?
JT65

E2D04 What technology is used to track, in real time, balloons carrying amateur radio transmitters?
APRS

E2D05 What is one advantage of the JT65 mode?
The ability to decode signals which have a very low signal-to-noise ratio

E2D06 Which of the following describes a method of establishing EME contacts?
Time synchronous transmissions alternately from each station

E2D07 What digital protocol is used by APRS?
AX.25

E2D08 What type of packet frame is used to transmit APRS beacon data?
Unnumbered Information

E2D09 What type of modulation is used for JT65 contacts?
Multi-tone AFSK

E2D10 How can an APRS station be used to help support a public service
An APRS station with a Global Positioning System unit can automatically transmit information to show a mobile station's position during the event.

E2D11 Which of the following data are used by the APRS network to communicate station location?
Latitude and longitude

SUMMARY – OPERATING PROCEDURES

E2E Operating Methods: Digital

E2E01 Which of the following types of modulation is common for data emissions below 30 MHz?
FSK

E2E02 What do the letters FEC mean as they relate to digital operation?
Forward Error Correction

E2E03 How is the timing of FT4 contacts organized?
Alternating transmissions at 7.5 second intervals

E2E04 What is indicated when one of the ellipses in an FSK crossed-ellipse display suddenly disappears?
Selective fading has occurred

E2E05 Which of these digital modes does not support keyboard-to-keyboard operation?
PACTOR

E2E06 What is the most common data rate used for HF packet?
300 baud

E2E07 Which of the following is a possible reason that attempts to initiate contact with a digital station on a clear frequency are unsuccessful?
Your transmit frequency is incorrect
The protocol version you are using is not supported by the digital station
Another station you are unable to hear is using the frequency
All these choices are correct

E2E08 Which of the following HF digital modes can be used to transfer binary files?
PACTOR

SUMMARY – OPERATING PROCEDURES

E2E09 Which of the following HF digital modes uses variable-length coding for bandwidth efficiency?
PSK31

E2E10 Which of these digital modes has the narrowest bandwidth?
PSK31

E2E11 What is the difference between direct FSK and audio FSK?
Direct FSK applies the data signal to the transmitter VFO, while AFSK transmits tones via phone

E2E12 How do ALE stations establish contact?
ALE constantly scans a list of frequencies, activating the radio when the designated call sign is received

E2E13 Which of these digital modes has the fastest data throughput under clear communication conditions?
300 baud packet

E3 - RADIO WAVE PROPAGATION

[3 Exam Questions - 3 Groups]

E3A Electromagnetic Waves

E3A01 What is the approximate maximum separation measured along the surface of the Earth between two stations communicating by EME?
12,000 miles, if the moon is visible by both stations

E3A02 What characterizes libration fading of an EME signal?
A fluttery irregular fading

E3A03 When scheduling EME contacts, which of these conditions will generally result in the least path loss?
When the moon is at perigee

E3A04 What do Hepburn maps predict?
Probability of tropospheric propagation

E3A05 Tropospheric propagation of microwave signals often occurs in association with what phenomenon?
Warm and cold fronts

E3A06 What might help to restore contact when DX signals become too weak to copy across an entire HF band a few hours after sunset?
Switch to a lower frequency HF band

E3A07 Atmospheric ducts capable of propagating microwave signals often form over what geographic feature?
Bodies of water

E3A08 When a meteor strikes the Earth's atmosphere, a cylindrical region of free electrons is formed at what layer of the ionosphere?
The E layer

SUMMARY – RADIO WAVE PROPAGATION

E3A09 Which of the following frequency ranges is most suited for meteor scatter communications?
28 MHz - 148 MHz

E3A10 Which type of atmospheric structure can create a path for microwave propagation?
Temperature inversion

E3A11 What is a typical range for tropospheric propagation of microwave signals?
100 miles to 300 miles

E3A12 What is the cause of auroral activity?
The interaction in the E layer of charged particles from the Sun with the Earth's magnetic field

E3A13 Which of these emission modes is best for auroral propagation?
CW

E3A14 What is meant by circularly polarized electromagnetic waves?
Waves with a rotating electric field

E3B Transequatorial Propagation

E3B01 What is transequatorial propagation?
Propagation between two mid-latitude points at approximately the same distance north and south of the magnetic equator

E3B02 What is the approximate maximum range for signals using transequatorial propagation?
5000 miles

E3B03 What is the best time of day for transequatorial propagation?
Afternoon or early evening

SUMMARY – RADIO WAVE PROPAGATION

E3B04 What is meant by the terms "extraordinary" and "ordinary" waves?
Independent waves created in the ionosphere that are elliptically polarized

E3B05 Which amateur bands typically support long-path propagation?
160 meters to 10 meters

E3B06 Which of the following amateur bands most frequently provides long-path propagation?
20 meters

E3B07 What happens to linearly polarized radio waves that split into ordinary and extraordinary waves in the ionosphere?
They become elliptically polarized

E3B08 Question withdrawn.

E3B09 At what time of year is sporadic E propagation most likely to occur?
Around the solstices, especially the summer solstice

E3B10 Why is chordal hop propagation desirable?
The signal experiences less loss compared to multi-hop using Earth as a reflector

E3B11 At what time of day can sporadic E propagation occur?
Any time

E3B12 What is the primary characteristic of chordal hop propagation?
Successive ionospheric refractions without an intermediate reflection from the ground

E3C Radio Propagation

E3C01 What does the radio communication term "ray tracing" describe?
Modeling a radio wave's path through the ionosphere

SUMMARY – RADIO WAVE PROPAGATION

E3C02 What is indicated by a rising A or K index?
Increasing disruption of the geomagnetic field

E3C03 Which of the following signal paths is most likely to experience high levels of absorption when the A index or K index is elevated?
Polar

E3C04 What does the value of Bz (B sub Z) represent?
Direction and strength of the interplanetary magnetic field

E3C05 What orientation of Bz (B sub z) increases the likelihood that incoming particles from the sun will cause disturbed conditions?
Southward

E3C06 By how much does the VHF/UHF radio horizon distance exceed the geometric horizon?
By approximately 15 percent of the distance

E3C07 Which of the following descriptors indicates the greatest solar flare intensity?
Class X

E3C08 What does the space weather term "G5" mean?
An extreme geomagnetic storm

E3C09 How does the intensity of an X3 flare compare to that of an X2 flare?
50 percent greater

E3C10 What does the 304A solar parameter measure?
UV emissions at 304 angstroms, correlated to the solar flux index

E3C11 What does VOACAP software model?
HF propagation

SUMMARY – RADIO WAVE PROPAGATION

E3C12 How does the maximum range of ground-wave propagation change when the signal frequency is increased?
It decreases

E3C13 What type of polarization is best for ground-wave propagation?
Vertical

E3C14 Why does the radio-path horizon distance exceed the geometric horizon?
Downward bending due to density variations in the atmosphere

E3C15 What might be indicated by a sudden rise in radio background noise across a large portion of the HF spectrum?
A solar flare has occurred

E4 - AMATEUR PRACTICES
[5 Exam Questions - 5 Groups]

E4A Test Equipment

E4A01 Which of the following limits the highest frequency signal that can be accurately displayed on a digital oscilloscope?
Sampling rate of the analog-to-digital converter

E4A02 Which of the following parameters does a spectrum analyzer display on the vertical and horizontal axes?
RF amplitude and frequency

E4A03 Which of the following test instruments is used to display spurious signals and/or intermodulation distortion products generated by an SSB transmitter?
A spectrum analyzer

E4A04 How is the compensation of an oscilloscope probe typically adjusted?
A square wave is displayed and the probe is adjusted until the horizontal portions of the displayed wave are as nearly flat as possible

E4A05 What is the purpose of the prescaler function on a frequency counter?
It divides a higher frequency signal so a low-frequency counter can display the input frequency

E4A06 What is the effect of aliasing on a digital oscilloscope caused by setting the time base too slow?
A false, jittery low-frequency version of the signal is displayed

SUMMARY – AMATEUR PRACTICES

E4A07 Which of the following is an advantage of using an antenna analyzer compared to an SWR bridge to measure antenna SWR?
Antenna analyzers do not need an external RF source

E4A08 Which of the following measures SWR?
An antenna analyzer

E4A09 Which of the following is good practice when using an oscilloscope probe?
Keep the signal ground connection of the probe as short as possible

E4A10 Which of the following displays multiple digital signal states simultaneously?
Logic analyzer

E4A11 How should an antenna analyzer be connected when measuring antenna resonance and feed point impedance?
Connect the antenna feed line directly to the analyzer's connector

E4B Measurements

E4B01 Which of the following factors most affects the accuracy of a frequency counter?
Time base accuracy

E4B02 What is the significance of voltmeter sensitivity expressed in ohms per volt?
The full scale reading of the voltmeter multiplied by its ohms per volt rating will indicate the input impedance of the voltmeter

E4B03 Which S parameter is equivalent to forward gain?
S21

SUMMARY – AMATEUR PRACTICES

E4B04 Which S parameter represents input port return loss or reflection coefficient (equivalent to VSWR)?
S11

E4B05 What three test loads are used to calibrate an RF vector network analyzer?
Short circuit, open circuit, and 50 ohms

E4B06 How much power is being absorbed by the load when a directional power meter connected between a transmitter and a terminating load reads 100 watts forward power and 25 watts reflected power?
75 watts

E4B07 What do the subscripts of S parameters represent?
The port or ports at which measurements are made

E4B08 Which of the following can be used to measure the Q of a series-tuned circuit?
The bandwidth of the circuit's frequency response

E4B09 What is indicated if the current reading on an RF ammeter placed in series with the antenna feed line of a transmitter increases as the transmitter is tuned to resonance?
There is more power going into the antenna

E4B10 Which of the following methods measures intermodulation distortion in an SSB transmitter?
Modulate the transmitter using two AF signals having non-harmonically related frequencies and observe the RF output with a spectrum analyzer

E4B11 Which of the following can be measured with a vector network analyzer?
Input impedance
Output impedance
Reflection coefficient
All these choices are correct

SUMMARY – AMATEUR PRACTICES

E4C Receiver Performance

E4C01 What is an effect of excessive phase noise in a receiver's local oscillator?
It can combine with strong signals on nearby frequencies to generate interference

E4C02 Which of the following receiver circuits can be effective in eliminating interference from strong out-of-band signals?
A front-end filter or pre-selector

E4C03 What is the term for the suppression in an FM receiver of one signal by another stronger signal on the same frequency?
Capture effect

E4C04 What is the noise figure of a receiver?
The ratio in dB of the noise generated by the receiver to the theoretical minimum noise

E4C05 What does a receiver noise floor of -174 dBm represent?
The theoretical noise in a 1 Hz bandwidth at the input of a perfect receiver at room temperature

E4C06 A CW receiver with the AGC off has an equivalent input noise power density of -174 dBm/Hz. What would be the level of an unmodulated carrier input to this receiver that would yield an audio output SNR of 0 dB in a 400 Hz noise bandwidth?
-148 dBm

E4C07 What does the MDS of a receiver represent?
The minimum discernible signal

E4C08 An SDR receiver is overloaded when input signals exceed what level?
The reference voltage of the analog-to-digital converter

SUMMARY – AMATEUR PRACTICES

E4C09 Which of the following choices is a good reason for selecting a high frequency for the design of the IF in a superheterodyne HF or VHF communications receiver?
Easier for front-end circuitry to eliminate image responses

E4C10 What is an advantage of having a variety of receiver IF bandwidths from which to select?
Receive bandwidth can be set to match the modulation bandwidth, maximizing signal-to-noise ratio and minimizing interference

E4C11 Why can an attenuator be used to reduce receiver overload on the lower frequency HF bands with little or no impact on signal-to-noise ratio?
Atmospheric noise is generally greater than internally generated noise even after attenuation

E4C12 Which of the following has the largest effect on an SDR receiver's dynamic range?
Analog-to-digital converter sample width in bits

E4C13 How does a narrow-band roofing filter affect receiver performance?
It improves dynamic range by attenuating strong signals near the receive frequency

E4C14 What transmit frequency might generate an image response signal in a receiver tuned to 14.300 MHz and that uses a 455 kHz IF frequency?
15.210 MHz

E4C15 What is reciprocal mixing?
Local oscillator phase noise mixing with adjacent strong signals to create interference to desired signals

SUMMARY – AMATEUR PRACTICES

E4D Receiver Performance

E4D01 What is meant by the blocking dynamic range of a receiver?
The difference in dB between the noise floor and the level of an incoming signal that will cause 1 dB of gain compression

E4D02 Which of the following describes problems caused by poor dynamic range in a receiver?
Spurious signals caused by cross-modulation and desensitization from strong adjacent signals

E4D03 How can intermodulation interference between two repeaters occur?
When the repeaters are in close proximity and the signals mix in the final amplifier of one or both transmitters

E4D04 Which of the following may reduce or eliminate intermodulation interference in a repeater caused by another transmitter operating in close proximity?
A properly terminated circulator at the output of the repeater's transmitter

E4D05 What transmitter frequencies would cause an intermodulation-product signal in a receiver tuned to 146.70 MHz when a nearby station transmits on 146.52 MHz?
146.34 MHz and 146.61 MHz

E4D06 What is the term for spurious signals generated by the combination of two or more signals in a non-linear device or circuit?
Intermodulation

E4D07 Which of the following reduces the likelihood of receiver desensitization?
Decrease the RF bandwidth of the receiver

SUMMARY - AMATEUR PRACTICES

E4D08 What causes intermodulation in an electronic circuit?
Nonlinear circuits or devices

E4D09 What is the purpose of the preselector in a communications receiver?
To increase rejection of signals outside the desired band

E4D10 What does a third-order intercept level of 40 dBm mean with respect to receiver performance?
A pair of 40 dBm input signals will theoretically generate a third-order intermodulation product that has the same output amplitude as either of the input signals

E4D11 Why are odd-order intermodulation products, created within a receiver, of particular interest compared to other products?
Odd-order products of two signals in the band of interest are also likely to be within the band

E4D12 What is the term for the reduction in receiver sensitivity caused by a strong signal near the received frequency?
Desensitization

E4E Noise Suppression and Grounding

E4E01 What problem can occur when using an automatic notch filter (ANF) to remove interfering carriers while receiving CW signals?
Removal of the CW signal as well as the interfering carrier

E4E02 Which of the following types of noise can often be reduced with a digital signal processing noise filter?
Broadband white noise
Ignition noise
Power line noise
All these choices are correct

SUMMARY – AMATEUR PRACTICES

E4E03 Which of the following signals might a receiver noise blanker be able to remove from desired signals?
Signals that appear across a wide bandwidth

E4E04 How can conducted and radiated noise caused by an automobile alternator be suppressed?
By connecting the radio's power leads directly to the battery and by installing coaxial capacitors in line with the alternator leads

E4E05 How can radio frequency interference from an AC motor be suppressed?
By installing a brute-force AC-line filter in series with the motor leads

E4E06 What is one type of electrical interference that might be caused by a nearby personal computer?
The appearance of unstable modulated or unmodulated signals at specific frequencies

E4E07 Which of the following can cause shielded cables to radiate or receive interference?
Common mode currents on the shield and conductors

E4E08 What current flows equally on all conductors of an unshielded multi-conductor cable?
Common-mode current

E4E09 What undesirable effect can occur when using an IF noise blanker?
Nearby signals may appear to be excessively wide even if they meet emission standards

SUMMARY – AMATEUR PRACTICES

E4E10 What might be the cause of a loud roaring or buzzing AC line interference that comes and goes at intervals?
Arcing contacts in a thermostatically controlled device
A defective doorbell or doorbell transformer inside a nearby residence
A malfunctioning illuminated advertising display
All these choices are correct

E4E11 What could cause local AM broadcast band signals to combine to generate spurious signals in the MF or HF bands?
Nearby corroded metal joints are mixing and re-radiating the broadcast signals

E5 – ELECTRICAL PRINCIPLES
[4 Exam Questions - 4 Groups]

E5A Resonance and Q

E5A01 What can cause the voltage across reactances in a series RLC circuit to be higher than the voltage applied to the entire circuit?
Resonance

E5A02 What is resonance in an LC or RLC circuit?
The frequency at which the capacitive reactance equals the inductive reactance

E5A03 What is the magnitude of the impedance of a series RLC circuit at resonance?
Approximately equal to circuit resistance

E5A04 What is the magnitude of the impedance of a parallel RLC circuit at resonance?
Approximately equal to circuit resistance

E5A05 What is the result of increasing the Q of an impedance-matching circuit?
Matching bandwidth is decreased

E5A06 What is the magnitude of the circulating current within the components of a parallel LC circuit at resonance?
It is at a maximum

E5A07 What is the magnitude of the current at the input of a parallel RLC circuit at resonance?
Minimum

E5A08 What is the phase relationship between the current through and the voltage across a series resonant circuit at resonance?
The voltage and current are in phase

SUMMARY – ELECTRICAL PRINCIPLES

E5A09 How is the Q of an RLC parallel resonant circuit calculated?
Resistance divided by the reactance of either the inductance or capacitance

E5A10 How is the Q of an RLC series resonant circuit calculated?
Reactance of either the inductance or capacitance divided by the resistance

E5A11 What is the half-power bandwidth of a resonant circuit that has a resonant frequency of 7.1 MHz and a Q of 150?
47.3 kHz

E5A12 What is the half-power bandwidth of a resonant circuit that has a resonant frequency of 3.7 MHz and a Q of 118?
31.4 kHz

E5A13 What is an effect of increasing Q in a series resonant circuit?
Internal voltages increase

E5A14 What is the resonant frequency of an RLC circuit if R is 22 ohms, L is 50 microhenries and C is 40 picofarads?
3.56 MHz

E5A15 Which of the following increases Q for inductors and capacitors?
Lower losses

E5A16 What is the resonant frequency of an RLC circuit if R is 33 ohms, L is 50 microhenries and C is 10 picofarads?
7.12 MHz

SUMMARY – ELECTRICAL PRINCIPLES

E5B Time and Phase

E5B01 What is the term for the time required for the capacitor in an RC circuit to be charged to 63.2% of the applied voltage or to discharge to 36.8% of its initial voltage?
One time constant."

E5B02 What letter is commonly used to represent susceptance?
Letter B

E5B03 How is impedance in polar form converted to an equivalent admittance?
Take the reciprocal of the magnitude and change the sign of the angle

E5B04 What is the time constant of a circuit having two 220-microfarad capacitors and two 1-megohm resistors, all in parallel?
220 seconds

E5B05 What happens to the magnitude of a pure reactance when it is converted to a susceptance?
It becomes the reciprocal

E5B06 What is susceptance?
The imaginary part of admittance

E5B07 What is the phase angle between the voltage across and the current through a series RLC circuit if XC is 500 ohms, R is 1 kilohm, and XL is 250 ohms?
14.0 degrees with the voltage lagging the current

E5B08 What is the phase angle between the voltage across and the current through a series RLC circuit if XC is 100 ohms, R is 100 ohms, and XL is 75 ohms?
14 degrees with the voltage lagging the current

SUMMARY – ELECTRICAL PRINCIPLES

E5B09 What is the relationship between the AC current through a capacitor and the voltage across a capacitor?
Current leads voltage by 90 degrees

E5B10 What is the relationship between the AC current through an inductor and the voltage across an inductor?
Voltage leads current by 90 degrees

E5B11 What is the phase angle between the voltage across and the current through a series RLC circuit if XC is 25 ohms, R is 100 ohms, and XL is 50 ohms?
14 degrees with the voltage leading the current

E5B12 What is admittance?
The inverse of impedance

E5C Coordinate Systems

E5C01 Which of the following represents capacitive reactance in rectangular notation?
–jX

E5C02 How are impedances described in polar coordinates?
By phase angle and magnitude

E5C03 Which of the following represents an inductive reactance in polar coordinates?
A positive phase angle

E5C04 What coordinate system is often used to display the resistive, inductive, and/or capacitive reactance components of impedance?
Rectangular coordinates

E5C05 What is the name of the diagram used to show the phase relationship between impedances at a given frequency?
Phasor diagram

SUMMARY - ELECTRICAL PRINCIPLES

E5C06 What does the impedance 50–j25 represent?
50 ohms resistance in series with 25 ohms capacitive reactance

E5C07 Where is the impedance of a pure resistance plotted on rectangular coordinates?
On the horizontal axis

E5C08 What coordinate system is often used to display the phase angle of a circuit containing resistance, inductive and/or capacitive reactance?
Polar coordinates

E5C09 When using rectangular coordinates to graph the impedance of a circuit, what do the axes represent?
The X axis represents the resistive component and the Y axis represents the reactive component

E5C10 Which point on Figure E5-1 best represents the impedance of a series circuit consisting of a 400-ohm resistor and a 38-picofarad capacitor at 14 MHz?
Point 4

E5C11 Which point in Figure E5-1 best represents the impedance of a series circuit consisting of a 300-ohm resistor and an 18-microhenry inductor at 3.505 MHz?
Point 3

E5C12 Which point on Figure E5-1 best represents the impedance of a series circuit consisting of a 300-ohm resistor and a 19-picofarad capacitor at 21.200 MHz?
Point 1

E5D AC and RF Energy

E5D01 What is the result of skin effect?
As frequency increases, RF current flows in a thinner layer of the conductor, closer to the surface

SUMMARY – ELECTRICAL PRINCIPLES

E5D02 Why is it important to keep lead lengths short for components used in circuits for VHF and above?
To avoid unwanted inductive reactance

E5D03 What is microstrip?
Precision printed circuit conductors above a ground plane that provide constant impedance interconnects at microwave frequencies

E5D04 Why are short connections used at microwave frequencies?
To reduce phase shift along the connection

E5D05 What is the power factor of an RL circuit having a 30-degree phase angle between the voltage and the current?
0.866

E5D06 In what direction is the magnetic field oriented about a conductor in relation to the direction of electron flow?
In a circle around the conductor

E5D07 How many watts are consumed in a circuit having a power factor of 0.71 if the apparent power is 500VA?
355 W

E5D08 How many watts are consumed in a circuit having a power factor of 0.6 if the input is 200VAC at 5 amperes?
600 watts

E5D09 What happens to reactive power in an AC circuit that has both ideal inductors and ideal capacitors?
It is repeatedly exchanged between the associated magnetic and electric fields, but is not dissipated

E5D10 How can the true power be determined in an AC circuit where the voltage and current are out of phase?
By multiplying the apparent power by the power factor

SUMMARY - ELECTRICAL PRINCIPLES

E5D11 What is the power factor of an RL circuit having a 60-degree phase angle between the voltage and the current?
0.5

E5D12 How many watts are consumed in a circuit having a power factor of 0.2 if the input is 100 VAC at 4 amperes?
80 watts

E5D13 How many watts are consumed in a circuit consisting of a 100-ohm resistor in series with a 100-ohm inductive reactance drawing 1 ampere?
100 watts

E5D14 What is reactive power?
Wattless, nonproductive power

E5D15 What is the power factor of an RL circuit having a 45-degree phase angle between the voltage and the current?
0.707

E6 - CIRCUIT COMPONENTS
6 Exam Questions - 6 Groups]

E6A Semiconductors

E6A01 In what application is gallium arsenide used as a semiconductor material?
In microwave circuits

E6A02 Which of the following semiconductor materials contains excess free electrons?
N-type

E6A03 Why does a PN-junction diode not conduct current when reverse biased?
Holes in P-type material and electrons in the N-type material are separated by the applied voltage, widening the depletion region

E6A04 What is the name given to an impurity atom that adds holes to a semiconductor crystal structure?
Acceptor impurity

E6A05 How does DC input impedance at the gate of a field-effect transistor compare with the DC input impedance of a bipolar transistor?
An FET has higher input impedance

E6A06 What is the beta of a bipolar junction transistor?
The change in collector current with respect to base current

E6A07 Which of the following indicates that a silicon NPN junction transistor is biased on?
Base-to-emitter voltage of approximately 0.6 to 0.7 volts

SUMMARY – CIRCUIT COMPONENTS

E6A08 What term indicates the frequency at which the grounded-base current gain of a transistor has decreased to 0.7 of the gain obtainable at 1 kHz?
Alpha cutoff frequency

E6A09 What is a depletion-mode FET?
An FET that exhibits a current flow between source and drain when no gate voltage is applied

E6A10 In Figure E6-1, (at the back of this book) what is the schematic symbol for an N-channel dual-gate MOSFET?
4

E6A11 In Figure E6-1, what is the schematic symbol for a P-channel junction FET?
1

E6A12 Why do many MOSFET devices have internally connected Zener diodes on the gates?
To reduce the chance of static damage to the gate

E6B Diodes

E6B01 What is the most useful characteristic of a Zener diode?
A constant voltage drop under conditions of varying current

E6B02 What is an important characteristic of a Schottky diode as compared to an ordinary silicon diode when used as a power supply rectifier?
Less forward voltage drop

E6B03 What type of bias is required for an LED to emit light?
Forward bias

E6B04 What type of semiconductor device is designed for use as a voltage-controlled capacitor?
Varactor diode

SUMMARY – CIRCUIT COMPONENTS

E6B05 What characteristic of a PIN diode makes it useful as an RF switch?
Low junction capacitance

E6B06 Which of the following is a common use of a Schottky diode?
As a VHF/UHF mixer or detector

E6B07 What is the failure mechanism when a junction diode fails due to excessive current?
Excessive junction temperature

E6B08 Which of the following is a Schottky barrier diode?
Metal-semiconductor junction

E6B09 What is a common use for point-contact diodes?
As an RF detector

E6B10 In Figure E6-2, (at the back of this book) what is the schematic symbol for a light-emitting diode?
5

E6B11 What is used to control the attenuation of RF signals by a PIN diode?
Forward DC bias current

E6C Digital ICs

E6C01 What is the function of hysteresis in a comparator?
To prevent input noise from causing unstable output signals

E6C02 What happens when the level of a comparator's input signal crosses the threshold?
The comparator changes its output state

E6C03 What is tri-state logic?
Logic devices with 0, 1, and high-impedance output states

SUMMARY – CIRCUIT COMPONENTS

E6C04 Which of the following is an advantage of BiCMOS logic?
It has the high input impedance of CMOS and the low output impedance of bipolar transistors

E6C05 What is an advantage of CMOS logic devices over TTL devices?
Lower power consumption

E6C06 Why do CMOS digital integrated circuits have high immunity to noise on the input signal or power supply?
The input switching threshold is about one-half the power supply voltage

E6C07 What best describes a pull-up or pull-down resistor?
A resistor connected to the positive or negative supply line used to establish a voltage when an input or output is an open circuit

E6C08 In Figure E6-3, what is the schematic symbol for a NAND gate?
2

E6C09 What is a Programmable Logic Device (PLD)?
A programmable collection of logic gates and circuits in a single integrated circuit

E6C10 In Figure E6-3, what is the schematic symbol for a NOR gate?
4

E6C11 In Figure E6-3, what is the schematic symbol for the NOT operation (inverter)?
5

SUMMARY – CIRCUIT COMPONENTS

E6D Toroidal and Solenoidal Inductors

E6D01 Why should core saturation of an impedance matching transformer be avoided?
Harmonics and distortion could result

E6D02 What is the equivalent circuit of a quartz crystal?
Motional capacitance, motional inductance, and loss resistance in series, all in parallel with a shunt capacitor representing electrode and stray capacitance

E6D03 Which of the following is an aspect of the piezoelectric effect?
Mechanical deformation of material by the application of a voltage

E6D04 Which materials are commonly used as a core in an inductor?
Ferrite and brass

E6D05 What is one reason for using ferrite cores rather than powdered iron in an inductor?
Ferrite toroids generally require fewer turns to produce a given inductance value

E6D06 What core material property determines the inductance of an inductor?
Permeability

E6D07 What is current in the primary winding of a transformer called if no load is attached to the secondary?
Magnetizing current

E6D08 What is one reason for using powdered-iron cores rather than ferrite cores in an inductor?
Powdered-iron cores generally maintain their characteristics at higher currents

Page 150   Extra Class – The Easy Way

SUMMARY – CIRCUIT COMPONENTS

E6D09 What devices are commonly used as VHF and UHF parasitic suppressors at the input and output terminals of a transistor HF amplifier?
Ferrite beads

E6D10 What is a primary advantage of using a toroidal core instead of a solenoidal core in an inductor?
Toroidal cores confine most of the magnetic field within the core material

E6D11 Which type of core material decreases inductance when inserted into a coil?
Brass

E6D12 What is inductor saturation?
The ability of the inductor's core to store magnetic energy has been exceeded

E6D13 What is the primary cause of inductor self-resonance?
Inter-turn capacitance

E6E Analog ICs: MMICs, IC packaging

E6E01 Why is gallium arsenide (GaAs) useful for semiconductor devices operating at UHF and higher frequencies?
Higher electron mobility

E6E02 Which of the following device packages is a through-hole type?
DIP

E6E03 Which of the following materials is likely to provide the highest frequency of operation when used in MMICs?
Gallium nitride

E6E04 Which is the most common input and output impedance of circuits that use MMICs?
50 ohms

SUMMARY – CIRCUIT COMPONENTS

E6E05 Which of the following noise figure values is typical of a low-noise UHF preamplifier?
2 dB

E6E06 What characteristics of the MMIC make it a popular choice for VHF through microwave circuits?
Controlled gain, low noise figure, and constant input and output impedance over the specified frequency range

E6E07 What type of transmission line is used for connections to MMICs?
Microstrip

E6E08 How is power supplied to the most common type of MMIC?
Through a resistor and/or RF choke connected to the amplifier output lead

E6E09 Which of the following component package types would be most suitable for use at frequencies above the HF range?
Surface mount

E6E10 What advantage does surface-mount technology offer at RF compared to using through-hole components?
Smaller circuit area
Shorter circuit-board traces
Components have less parasitic inductance and capacitance
All these choices are correct

E6E11 What is a characteristic of DIP packaging used for integrated circuits?
A total of two rows of connecting pins placed on opposite sides of the package (Dual In-line Package)

E6E12 Why are DIP through-hole package ICs not typically used at UHF and higher frequencies?
Excessive lead length

SUMMARY – CIRCUIT COMPONENTS

E6F Optical Components

E6F01 What absorbs the energy from light falling on a photovoltaic cell?
Electrons

E6F02 What happens to the conductivity of a photoconductive material when light shines on it?
It increases

E6F03 What is the most common configuration of an optoisolator or optocoupler?
An LED and a phototransistor

E6F04 What is the photovoltaic effect?
The conversion of light to electrical energy

E6F05 Which describes an optical shaft encoder?
A device that detects rotation of a control by interrupting a light source with a patterned wheel

E6F06 Which of these materials is most commonly used to create photoconductive devices?
A crystalline semiconductor

E6F07 What is a solid-state relay?
A device that uses semiconductors to implement the functions of an electromechanical relay

E6F08 Why are optoisolators often used in conjunction with solid-state circuits when switching 120 VAC?
Optoisolators provide a very high degree of electrical isolation between a control circuit and the circuit being switched

E6F09 What is the efficiency of a photovoltaic cell?
The relative fraction of light that is converted to current

SUMMARY – CIRCUIT COMPONENTS

E6F10 What is the most common type of photovoltaic cell used for electrical power generation?
Silicon

E6F11 What is the approximate open-circuit voltage produced by a fully illuminated silicon photovoltaic cell?
0.5 V

E7 – PRACTICAL CIRCUITS

[8 Exam Questions - 8 Groups]

E7A Digital Circuits

E7A01 Which circuit is bistable?
A flip-flop

E7A02 What is the function of a decade counter?
It produces one output pulse for every 10 input pulses

E7A03 Which of the following can divide the frequency of a pulse train by 2?
A flip-flop

E7A04 How many flip-flops are required to divide a signal frequency by 4?
2

E7A05 Which of the following is a circuit that continuously alternates between two states without an external clock?
Astable multivibrator

E7A06 What is a characteristic of a monostable multivibrator?
It switches momentarily to the opposite binary state and then returns to its original state after a set time

E7A07 What logical operation does a NAND gate perform?
It produces logic 0 at its output only when all inputs are logic 1

E7A08 What logical operation does an OR gate perform?
It produces logic 1 at its output if any or all inputs are logic 1

E7A09 What logical operation is performed by an exclusive NOR gate?
It produces logic 0 at its output if only one input is logic 1

SUMMARY – PRACTICAL CIRCUITS

E7A10 What is a truth table?
A list of inputs and corresponding outputs for a digital device

E7A11 What type of logic defines "1" as a high voltage?
Positive Logic

E7B Amplifiers

E7B01 For what portion of the signal cycle does each active element in a push-pull Class AB amplifier conduct?
More than 180 degrees but less than 360 degrees

E7B02 What is a Class D amplifier?
A type of amplifier that uses switching technology to achieve high efficiency

E7B03 Which of the following components form the output of a class D amplifier circuit?
A low-pass filter to remove switching signal components

E7B04 Where on the load line of a Class A common emitter amplifier would bias normally be set?
Approximately halfway between saturation and cutoff

E7B05 What can be done to prevent unwanted oscillations in an RF power amplifier?
Install parasitic suppressors and/or neutralize the stage

E7B06 Which of the following amplifier types reduces even-order harmonics?
Push-pull

E7B07 Which of the following is a likely result when a Class C amplifier is used to amplify a single-sideband phone signal?
Signal distortion and excessive bandwidth

E7B08 How can an RF power amplifier be neutralized?
By feeding a 180-degree out-of-phase portion of the output back to the input

SUMMARY – PRACTICAL CIRCUITS

E7B09 Which of the following describes how the loading and tuning capacitors are to be adjusted when tuning a vacuum tube RF power amplifier that employs a Pi-network output circuit?
The tuning capacitor is adjusted for minimum plate current, and the loading capacitor is adjusted for maximum permissible plate current

E7B10 In Figure E7-1, what is the purpose of R1 and R2?
Voltage divider bias

E7B11 In Figure E7-1, what is the purpose of R3?
Self bias

E7B12 What type of amplifier circuit is in Figure E7-1?
Common emitter

E7B13 Which of the following describes an emitter follower (or common collector) amplifier?
An amplifier with a low impedance output that follows the base input voltage

E7B14 Why are switching amplifiers more efficient than linear amplifiers?
The power transistor is at saturation or cutoff most of the time

E7B15 What is one way to prevent thermal runaway in a bipolar transistor amplifier?
Use a resistor in series with the emitter

E7B16 What is the effect of intermodulation products in a linear power amplifier?
Transmission of spurious signals

E7B17 Why are odd-order rather than even-order intermodulation distortion products of concern in linear power amplifiers?
Because they are relatively close in frequency to the desired signal

SUMMARY – PRACTICAL CIRCUITS

E7B18 What is a characteristic of a grounded-grid amplifier?
Low input impedance

E7C Filters and Matching Networks

E7C01 How are the capacitors and inductors of a low-pass filter Pi-network arranged between the network's input and output?
A capacitor is connected between the input and ground, another capacitor is connected between the output and ground, and an inductor is connected between input and output

E7C02 Which of the following is a property of a T-network with series capacitors and a parallel shunt inductor?
It is a high-pass filter

E7C03 What advantage does a series-L Pi-L network have over a series-L Pi-network for impedance matching between the final amplifier of a vacuum-tube transmitter and an antenna?
Greater harmonic suppression

E7C04 How does an impedance-matching circuit transform a complex impedance to a resistive impedance?
It cancels the reactive part of the impedance and changes the resistive part to a desired value

E7C05 Which filter type is described as having ripple in the passband and a sharp cutoff?
A Chebyshev filter

E7C06 What are the distinguishing features of an elliptical filter?
Extremely sharp cutoff with one or more notches in the stop band

E7C07 Which describes a Pi-L network used for matching a vacuum tube final amplifier to a 50-ohm unbalanced output?

SUMMARY – PRACTICAL CIRCUITS

A Pi-network with an additional series inductor on the output

E7C08 Which of the following factors has the greatest effect on the bandwidth and response shape of a crystal ladder filter?
The relative frequencies of the individual crystals

E7C09 What is a crystal lattice filter?
A filter with narrow bandwidth and steep skirts made using quartz crystals

E7C10 Which of the following filters would be the best choice for use in a 2 meter band repeater duplexer?
A cavity filter

E7C11 Which of the following describes a receiving filter's ability to reject signals occupying an adjacent channel?
Shape factor

E7C12 What is one advantage of a Pi-matching network over an L-matching network consisting of a single inductor and a single capacitor?
The Q of Pi-networks can be controlled

E7D Power Supplies

E7D01 How does a linear electronic voltage regulator work?
The conduction of a control element is varied to maintain a constant output voltage

E7D02 What is a characteristic of a switching electronic voltage regulator?
The controlled device's duty cycle is changed to produce a constant average output voltage

SUMMARY – PRACTICAL CIRCUITS

E7D03 What device is typically used as a stable voltage reference in a linear voltage regulator?
A Zener diode

E7D04 Which of the following types of linear voltage regulator usually make the most efficient use of the primary power source?
A series regulator

E7D05 Which of the following types of linear voltage regulator places a constant load on the unregulated voltage source?
A shunt regulator

E7D06 What is the purpose of Q1 in the circuit shown in Figure E7-2?
It controls the current supplied to the load

E7D07 What is the purpose of C2 in the circuit shown in Figure E7-2?
It bypasses rectifier output ripple around D1

E7D08 What type of circuit is shown in Figure E7-2?
Linear voltage regulator

E7D09 What is the main reason to use a charge controller with a solar power system?
Prevention of battery damage due to overcharge

E7D10 What is the primary reason that a high-frequency switching type high-voltage power supply can be both less expensive and lighter in weight than a conventional power supply?
The high frequency inverter design uses much smaller transformers and filter components for an equivalent power output

SUMMARY – PRACTICAL CIRCUITS

E7D11 What is the function of the pass transistor in a linear voltage regulator circuit?
Maintains nearly constant output voltage over a wide range of load current

E7D12 What is the drop-out voltage of an analog voltage regulator?
Minimum input-to-output voltage required to maintain regulation

E7D13 What is the equation for calculating power dissipated by a series linear voltage regulator?
Voltage difference from input to output multiplied by output current

E7D14 What is the purpose of connecting equal-value resistors across power supply filter capacitors connected in series?
Equalize the voltage across each capacitor
Discharge the capacitors when voltage is removed
Provide a minimum load on the supply
All these choices are correct

E7D15 What is the purpose of a step-start circuit in a high-voltage power supply?
To allow the filter capacitors to charge gradually

E7E Modulation and Demodulation

E7E01 Which of the following can be used to generate FM phone emissions?
A reactance modulator on the oscillator

E7E02 What is the function of a reactance modulator?
To produce PM or FM signals by using an electrically variable inductance or capacitance

SUMMARY – PRACTICAL CIRCUITS

E7E03 What is a frequency discriminator stage in a FM receiver?
A circuit for detecting FM signals

E7E04 What is one way a single-sideband phone signal can be generated?
By using a balanced modulator followed by a filter

E7E05 What circuit is added to an FM transmitter to boost the higher audio frequencies?
A pre-emphasis network

E7E06 Why is de-emphasis commonly used in FM communications receivers?
For compatibility with transmitters using phase modulation

E7E07 What is meant by the term "baseband" in radio communications?
The frequency range occupied by a message signal prior to modulation

E7E08 What are the principal frequencies that appear at the output of a mixer circuit?
The two input frequencies along with their sum and difference frequencies

E7E09 What occurs when an excessive amount of signal energy reaches a mixer circuit?
Spurious mixer products are generated

E7E10 How does a diode envelope detector function?
By rectification and filtering of RF signals

E7E11 Which type of detector circuit is used for demodulating SSB signals?
Product detector.

SUMMARY – PRACTICAL CIRCUITS

E7F DSP and Software Defined Radio

E7F01 What is meant by direct digital conversion as applied to software defined radios?
Incoming RF is digitized by an analog-to-digital converter without being mixed with a local oscillator signal

E7F02 What kind of digital signal processing audio filter is used to remove unwanted noise from a received SSB signal?
An adaptive filter

E7F03 What type of digital signal processing filter is used to generate an SSB signal?
A Hilbert-transform filter

E7F04 What is a common method of generating an SSB signal using digital signal processing?
Signals are combined in quadrature phase relationship

E7F05 How frequently must an analog signal be sampled by an analog-to-digital converter so that the signal can be accurately reproduced?
At least twice the rate of the highest frequency component of the signal

E7F06 What is the minimum number of bits required for an analog-to-digital converter to sample a signal with a range of 1 volt at a resolution of 1 millivolt?
10 bits

E7F07 What function is performed by a Fast Fourier Transform?
Converting digital signals from the time domain to the frequency domain

E7F08 What is the function of decimation?
Reducing the effective sample rate by removing samples

SUMMARY – PRACTICAL CIRCUITS

E7F09 Why is an anti-aliasing digital filter required in a digital decimator?
It removes high-frequency signal components that would otherwise be reproduced as lower frequency components

E7F10 What aspect of receiver analog-to-digital conversion determines the maximum receive bandwidth of a Direct Digital Conversion SDR?
Sample rate

E7F11 What sets the minimum detectable signal level for a direct-sampling SDR receiver in the absence of atmospheric or thermal noise?
Reference voltage level and sample width in bits

E7F12 Which of the following is an advantage of a Finite Impulse Response (FIR) filter vs an Infinite Impulse Response (IIR) digital filter?
FIR filters can delay all frequency components of the signal by the same amount

E7F13 What is the function of taps in a digital signal processing filter?
Provide incremental signal delays for filter algorithms

E7F14 Which of the following would allow a digital signal processing filter to create a sharper filter response?
More taps

E7G Active filters and Op-amp Circuits

E7G01 What is the typical output impedance of an op-amp?
Very low

E7G02 What is ringing in a filter?
Undesired oscillations added to the desired signal

E7G03 What is the typical input impedance of an op-amp?
Very high

SUMMARY – PRACTICAL CIRCUITS

E7G04 What is meant by the term "op-amp input offset voltage"?
The differential input voltage needed to bring the open loop output voltage to zero

E7G05 How can unwanted ringing and audio instability be prevented in an op-amp RC audio filter circuit?
Restrict both gain and Q

E7G06 What is the gain-bandwidth of an operational amplifier?
The frequency at which the open-loop gain of the amplifier equals one

E7G07 What magnitude of voltage gain can be expected from the circuit in Figure E7-3 when R1 is 10 ohms and RF is 470 ohms?
47

E7G08 How does the gain of an ideal operational amplifier vary with frequency?
It does not vary with frequency

E7G09 What will be the output voltage of the circuit shown in Figure E7-3 if R1 is 1000 ohms, RF is 10,000 ohms, and 0.23 volts DC is applied to the input?
-2.3 volts

E7G10 What absolute voltage gain can be expected from the circuit in Figure E7-3 when R1 is 1800 ohms and RF is 68 kilohms?
38

E7G11 What absolute voltage gain can be expected from the circuit in Figure E7-3 when R1 is 3300 ohms and RF is 47 kilohms?
14

SUMMARY – PRACTICAL CIRCUITS

E7G12 What is an operational amplifier?
A high-gain, direct-coupled differential amplifier with very high input impedance and very low output impedance

E7H Oscillators and Signal Sources

E7H01 What are three oscillator circuits used in amateur radio equipment?
Colpitts, Hartley and Pierce

E7H02 What is a microphonic?
Changes in oscillator frequency due to mechanical vibration

E7H03 How is positive feedback supplied in a Hartley oscillator?
Through a tapped coil

E7H04 How is positive feedback supplied in a Colpitts oscillator?
Through a capacitive divider

E7H05 How is positive feedback supplied in a Pierce oscillator?
Through a quartz crystal

E7H06 Which of the following oscillator circuits are commonly used in VFOs?
Colpitts and Hartley

E7H07 How can an oscillator's microphonic responses be reduced?
Mechanically isolate the oscillator circuitry from its enclosure

E7H08 Which of the following components can be used to reduce thermal drift in crystal oscillators?
NP0 capacitors

SUMMARY – PRACTICAL CIRCUITS

E7H09 What type of frequency synthesizer circuit uses a phase accumulator, lookup table, digital to analog converter, and a low-pass anti-alias filter?
A direct digital synthesizer

E7H10 What information is contained in the lookup table of a direct digital synthesizer (DDS)?
Amplitude values that represent the desired waveform

E7H11 What are the major spectral impurity components of direct digital synthesizers?
Spurious signals at discrete frequencies

E7H12 Which of the following must be done to ensure that a crystal oscillator provides the frequency specified by the crystal manufacturer?
Provide the crystal with a specified parallel capacitance

E7H13 Which of the following is a technique for providing highly accurate and stable oscillators needed for microwave transmission and reception?
Use a GPS signal reference
Use a rubidium stabilized reference oscillator
Use a temperature-controlled high Q dielectric resonator
All these choices are correct

E7H14 What is a phase-locked loop circuit?
An electronic servo loop consisting of a phase detector, a low-pass filter, a voltage-controlled oscillator, and a stable reference oscillator

E7H15 Which of these functions can be performed by a phase-locked loop?
Frequency synthesis, FM demodulation

E8 - SIGNALS AND EMISSIONS
4 Exam Questions - 4 Groups]

E8A Waveforms

E8A01 What is the name of the process that shows that a square wave is made up of a sine wave plus all its odd harmonics?
Fourier analysis

E8A02 Which of the following is a type of analog-to-digital conversion?
Successive approximation

E8A03 What type of wave does a Fourier analysis show to be made up of sine waves of a given fundamental frequency plus all its harmonics?
A sawtooth wave

E8A04 What is "dither" with respect to analog-to-digital converters?
A small amount of noise added to the input signal to allow more precise representation of a signal over time

E8A05 What of the following instruments would be the most accurate for measuring the RMS voltage of a complex waveform?
A true-RMS calculating meter

E8A06 What is the approximate ratio of PEP-to-average power in a typical single-sideband phone signal?
2.5 to 1

E8A07 What determines the PEP-to-average power ratio of a single-sideband phone signal?
Speech characteristics

SUMMARY – SIGNALS AND EMISSIONS

E8A08 Why would a direct or flash conversion analog-to-digital converter be useful for a software defined radio?
Very high speed allows digitizing high frequencies

E8A09 How many different input levels can be encoded by an analog-to-digital converter with 8-bit resolution?
256

E8A10 What is the purpose of a low-pass filter used in conjunction with a digital-to-analog converter?
Remove harmonics from the output caused by the discrete analog levels generated

E8A11 Which of the following is a measure of the quality of an analog-to-digital converter
Total harmonic distortion

E8B Modulation and Demodulation

E8B01 What is the modulation index of an FM signal?
The ratio of frequency deviation to modulating signal frequency

E8B02 How does the modulation index of a phase-modulated emission vary with RF carrier frequency?
It does not depend on the RF carrier frequency

E8B03 What is the modulation index of an FM-phone signal having a maximum frequency deviation of 3000 Hz either side of the carrier frequency when the modulating frequency is 1000 Hz?
3

E8B04 What is the modulation index of an FM-phone signal having a maximum carrier deviation of plus or minus 6 kHz when modulated with a 2 kHz modulating frequency?
3

SUMMARY - SIGNALS AND EMISSIONS

E8B05 What is the deviation ratio of an FM-phone signal having a maximum frequency swing of plus-or-minus 5 kHz when the maximum modulation frequency is 3 kHz?
1.67

E8B06 What is the deviation ratio of an FM-phone signal having a maximum frequency swing of plus or minus 7.5 kHz when the maximum modulation frequency is 3.5 kHz?
2.14

E8B07 Orthogonal Frequency Division Multiplexing is a technique used for which type of amateur communication?
High-speed digital modes

E8B08 What describes Orthogonal Frequency Division Multiplexing?
A digital modulation technique using subcarriers at frequencies chosen to avoid intersymbol interference

E8B09 What is deviation ratio?
The ratio of the maximum carrier frequency deviation to the highest audio modulating frequency

E8B10 What is frequency division multiplexing?
Two or more information streams are merged into a baseband, which then modulates the transmitter

E8B11 What is digital time division multiplexing?
Two or more signals are arranged to share discrete time slots of a data transmission

E8C Digital Signals

E8C01 How is Forward Error Correction implemented?
By transmitting extra data that may be used to detect and correct transmission errors

E8C02 What is the definition of symbol rate in a digital transmission?

SUMMARY – SIGNALS AND EMISSIONS

The rate at which the waveform changes to convey information

E8C03 Why should phase-shifting of a PSK signal be done at the zero crossing of the RF signal?
To minimize bandwidth

E8C04 What technique minimizes the bandwidth of a PSK31 signal?
Use of sinusoidal data pulses

E8C05 What is the approximate bandwidth of a 13-WPM International Morse Code transmission?
52 Hz

E8C06 What is the bandwidth of a 170-hertz shift, 300-baud ASCII transmission?
0.5 kHz

E8C07 What is the bandwidth of a 4800-Hz frequency shift, 9600-baud ASCII FM transmission?
15.36 kHz

E8C08 How does ARQ accomplish error correction?
If errors are detected, a retransmission is requested

E8C09 Which digital code allows only one bit to change between sequential code values?
Gray code

E8C10 How may data rate be increased without increasing bandwidth?
Using a more efficient digital code

E8C11 What is the relationship between symbol rate and baud?
They are the same

SUMMARY – SIGNALS AND EMISSIONS

E8C12 What factors affect the bandwidth of a transmitted CW signal?
Keying speed and shape factor (rise and fall time)

E8D Keying defects and Overmodulation

E8D01 Why are received spread spectrum signals resistant to interference?
Signals not using the spread spectrum algorithm are suppressed in the receiver

E8D02 What spread spectrum communications technique uses a high-speed binary bit stream to shift the phase of an RF carrier?
Direct sequence

E8D03 How does the spread spectrum technique of frequency hopping work?
The frequency of the transmitted signal is changed very rapidly according to a pseudorandom sequence also used by the receiving station

E8D04 What is the primary effect of extremely short rise or fall time on a CW signal?
The generation of key clicks

E8D05 What is the most common method of reducing key clicks?
Increase keying waveform rise and fall times

E8D06 What is the advantage of including parity bits in ASCII characters?
Some types of errors can be detected

E8D07 What is a common cause of overmodulation of AFSK signals?
Excessive transmit audio levels

Page 172   Extra Class - The Easy Way

SUMMARY – SIGNALS AND EMISSIONS

E8D08 What parameter evaluates distortion of an AFSK signal caused by excessive input audio levels?
Intermodulation Distortion (IMD)

E8D09 What is considered an acceptable maximum IMD level for an idling PSK signal?
-30 dB

E8D10 What are some of the differences between the Baudot digital code and ASCII?
Baudot uses 5 data bits per character, ASCII uses 7 or 8; Baudot uses 2 characters as letters/figures shift codes, ASCII has no letters/figures shift code

E8D11 What is one advantage of using ASCII code for data communications?
It is possible to transmit both upper and lower case text

E9 - ANTENNAS AND TRANSMISSION LINES

[8 Exam Questions - 8 Groups]

E9A Basic Antenna Parameters

E9A01 What is an isotropic antenna?
A theoretical, omnidirectional antenna used as a reference for antenna gain

E9A02 What is the effective radiated power relative to a dipole of a repeater station with 150 watts transmitter power output, 2 dB feed line loss, 2.2 dB duplexer loss, and 7 dBd antenna gain?
286 watts

E9A03 What is the radiation resistance of an antenna?
The value of a resistance that would dissipate the same amount of power as that radiated from an antenna

E9A04 Which of the following factors affect the feed point impedance of an antenna?
Antenna height

E9A05 What is included in the total resistance of an antenna system?
Radiation resistance plus loss resistance

E9A06 What is the effective radiated power relative to a dipole of a repeater station with 200 watts transmitter power output, 4 dB feed line loss, 3.2 dB duplexer loss, 0.8 dB circulator loss, and 10 dBd antenna gain?
317 watts

SUMMARY – ANTENNAS AND TRANSMISSION LINES

E9A07 What is the effective isotropic radiated power of a repeater station with 200 watts transmitter power output, 2 dB feed line loss, 2.8 dB duplexer loss, 1.2 dB circulator loss, and 7 dBi antenna gain?
252 watts

E9A08 What is antenna bandwidth?
The frequency range over which an antenna satisfies a performance requirement

E9A09 What is antenna efficiency?
Radiation resistance divided by total resistance

E9A10 Which of the following improves the efficiency of a ground-mounted quarter-wave vertical antenna?
Installing a radial system

E9A11 Which of the following factors determines ground losses for a ground-mounted vertical antenna operating in the 3 MHz to 30 MHz range?
Soil conductivity

E9A12 How much gain does an antenna have compared to a 1/2-wavelength dipole when it has 6 dB gain over an isotropic antenna?
3.85 dB

E9A13 What term describes station output, taking into account all gains and losses?
Effective radiated power

E9B Antenna Patterns

E9B01 In the antenna radiation pattern shown in Figure E9-1, what is the beamwidth?
50 degrees

SUMMARY – ANTENNAS AND TRANSMISSION LINES

E9B02 In the antenna radiation pattern shown in Figure E9-1, what is the front-to-back ratio?
18 dB

E9B03 In the antenna radiation pattern shown in Figure E9-1, what is the front-to-side ratio?
14 dB

E9B04 What is the front-to-back ratio of the radiation pattern shown in Figure E9-2?
28 dB

E9B05 What type of antenna pattern is shown in Figure E9-2?
Elevation

E9B06 What is the elevation angle of peak response in the antenna radiation pattern shown in Figure E9-2?
7.5 degrees

E9B07 How does the total amount of radiation emitted by a directional gain antenna compare with the total amount of radiation emitted from a theoretical isotropic antenna, assuming each is driven by the same amount of power?
They are the same

E9B08 What is the far field of an antenna?
The region where the shape of the antenna pattern is independent of distance

E9B09 What type of computer program technique is commonly used for modeling antennas?
Method of Moments

E9B10 What is the principle of a Method of Moments analysis?
A wire is modeled as a series of segments, each having a uniform value of current

SUMMARY – ANTENNAS AND TRANSMISSION LINES

E9B11 What is a disadvantage of decreasing the number of wire segments in an antenna model below 10 segments per half-wavelength?
The computed feed point impedance may be incorrect

E9C Wire and Phased Array Antennas

E9C01 What is the radiation pattern of two 1/4-wavelength vertical antennas spaced 1/2-wavelength apart and fed 180 degrees out of phase?
A figure-8 oriented along the axis of the array

E9C02 What is the radiation pattern of two 1/4-wavelength vertical antennas spaced 1/4 wavelength apart and fed 90 degrees out of phase?
Cardioid

E9C03 What is the radiation pattern of two 1/4-wavelength vertical antennas spaced 1/2-wavelength apart and fed in phase?
A Figure-8 broadside to the axis of the array

E9C04 What happens to the radiation pattern of an unterminated long wire antenna as the wire length is increased?
The lobes align more in the direction of the wire

E9C05 Which of the following is a type of OCFD antenna?
A dipole fed approximately 1/3 the way from one end with a 4:1 balun to provide multiband operation

E9C06 What is the effect of adding a terminating resistor to a rhombic antenna?
It changes the radiation pattern from bidirectional to unidirectional

E9C07 What is the approximate feed point impedance at the center of a two-wire folded dipole antenna?
300 ohms

SUMMARY – ANTENNAS AND TRANSMISSION LINES

E9C08 What is a folded dipole antenna?
A half-wave dipole with an additional parallel wire connecting its two ends

E9C09 Which of the following describes a G5RV antenna?
A multi-band dipole antenna fed with coax and a balun through a selected length of open wire transmission line

E9C10 Which of the following describes a Zepp antenna?
An end-fed dipole antenna

E9C11 How is the far-field elevation pattern of a vertically polarized antenna affected by being mounted over seawater versus soil?
The low-angle radiation increases

E9C12 Which of the following describes an Extended Double Zepp antenna?
A center-fed 1.25-wavelength antenna (two 5/8-wave elements in phase)

E9C13 How does the radiation pattern of a horizontally polarized 3-element beam antenna vary with increasing height above ground?
The takeoff angle of the lowest elevation lobe decreases

E9C14 How does the performance of a horizontally polarized antenna mounted on the side of a hill compare with the same antenna mounted on flat ground?
The main lobe takeoff angle decreases in the downhill direction

E9D Directional and Short Antennas

E9D01 How much does the gain of an ideal parabolic dish antenna change when the operating frequency is doubled?
6 dB

SUMMARY – ANTENNAS AND TRANSMISSION LINES

E9D02 How can linearly polarized Yagi antennas be used to produce circular polarization?
Arrange two Yagis perpendicular to each other with the driven elements at the same point on the boom fed 90 degrees out of phase

E9D03 Where should a high Q loading coil be placed to minimize losses in a shortened vertical antenna?
Near the center of the vertical radiator

E9D04 Why should an HF mobile antenna loading coil have a high ratio of reactance to resistance?
To minimize losses

E9D05 What usually occurs if a Yagi antenna is designed solely for maximum forward gain?
The front-to-back ratio decreases

E9D06 What happens to the SWR bandwidth when one or more loading coils are used to resonate an electrically short antenna?
It is decreased

E9D07 What is an advantage of using top loading in a shortened HF vertical antenna?
Improved radiation efficiency

E9D08 What happens as the Q of an antenna increases?
SWR bandwidth decreases

E9D09 What is the function of a loading coil used as part of an HF mobile antenna?
To cancel capacitive reactance

SUMMARY – ANTENNAS AND TRANSMISSION LINES

E9D10 What happens to feed-point impedance at the base of a fixed length HF mobile antenna when operated below its resonant frequency?
The radiation resistance decreases and the capacitive reactance increases

E9D11 Which of the following conductors would be best for minimizing losses in a station's RF ground system?
Wide flat copper strap

E9D12 Which of the following would provide the best RF ground for your station?
An electrically short connection to 3 or 4 interconnected ground rods driven into the Earth

E9E Matching

E9E01 What system matches a higher-impedance transmission line to a lower-impedance antenna by connecting the line to the driven element in two places spaced a fraction of a wavelength each side of element center?
The delta matching system

E9E02 What is the name of an antenna matching system that matches an unbalanced feed line to an antenna by feeding the driven element both at the center of the element and at a fraction of a wavelength to one side of center?
The gamma match

E9E03 What is the name of the matching system that uses a section of transmission line connected in parallel with the feed line at or near the feed point?
The stub match

E9E04 What is the purpose of the series capacitor in a gamma-type antenna matching network?
To cancel the inductive reactance of the matching network

SUMMARY – ANTENNAS AND TRANSMISSION LINES

E9E05 How must an antenna's driven element be tuned to use a hairpin matching system?
The driven element reactance must be capacitive

E9E06 Which of these feed line impedances would be suitable for constructing a quarter-wave Q-section for matching a 100-ohm loop to 50-ohm feed line?
75 ohms

E9E07 What parameter describes the interactions at the load end of a mismatched transmission line?
Reflection coefficient

E9E08 What is a use for a Wilkinson divider?
It is used to divide power equally between two 50-ohm loads while maintaining 50-ohm input impedance

E9E09 Which of the following is used to shunt-feed a grounded tower at its base?
Gamma match

E9E10 Which of these choices is an effective way to match an antenna with a 100-ohm feed point impedance to a 50-ohm coaxial cable feed line?
Insert a 1/4-wavelength piece of 75-ohm coaxial cable transmission line in series between the antenna terminals and the 50-ohm feed cable

E9E11 What is the primary purpose of phasing lines when used with an antenna having multiple driven elements?
It ensures that each driven element operates in concert with the others to create the desired antenna pattern

E9F Transmission Lines

E9F01 What is the velocity factor of a transmission line?
The velocity of the wave in the transmission line divided by the velocity of light in a vacuum

SUMMARY – ANTENNAS AND TRANSMISSION LINES

E9F02 Which of the following has the biggest effect on the velocity factor of a transmission line?
Dielectric materials used in the line

E9F03 Why is the physical length of a coaxial cable transmission line shorter than its electrical length?
Electrical signals move more slowly in a coaxial cable than in air

E9F04 What impedance does a 1/2-wavelength transmission line present to a generator when the line is shorted at the far end?
Very low impedance

E9F05 What is the approximate physical length of a solid polyethylene dielectric coaxial transmission line that is electrically 1/4 wavelength long at 14.1 MHz?
3.5 meters

E9F06 What is the approximate physical length of an air-insulated, parallel conductor transmission line that is electrically 1/2 wavelength long at 14.10 MHz?
10.6 meters

E9F07 How does ladder line compare to small-diameter coaxial cable such as RG-58 at 50 MHz?
Lower loss

E9F08 Which of the following is a significant difference between foam dielectric coaxial cable and solid dielectric cable, assuming all other parameters are the same?
Foam dielectric has lower safe operating voltage limits
Foam dielectric has lower loss per unit of length
Foam dielectric has higher velocity factor
All these choices are correct

SUMMARY – ANTENNAS AND TRANSMISSION LINES

E9F09 What is the approximate physical length of a foam polyethylene dielectric coaxial transmission line that is electrically 1/4 wavelength long at 7.2 MHz?
8.3 meters

E9F10 What impedance does a 1/8-wavelength transmission line present to a generator when the line is shorted at the far end?
An inductive reactance

E9F11 What impedance does a 1/8-wavelength transmission line present to a generator when the line is open at the far end?
A capacitive reactance

E9F12 What impedance does a 1/4-wavelength transmission line present to a generator when the line is open at the far end?
Very low impedance

E9F13 What impedance does a 1/4-wavelength transmission line present to a generator when the line is shorted at the far end?
Very high impedance

E9G Smith Chart

E9G01 Which of the following can be calculated using a Smith chart?
Impedance along transmission lines

E9G02 What type of coordinate system is used in a Smith chart?
Resistance circles and reactance arcs

E9G03 Which of the following is often determined using a Smith chart?
Impedance and SWR values in transmission lines

SUMMARY – ANTENNAS AND TRANSMISSION LINES

E9G04 What are the two families of circles and arcs that make up a Smith chart?
Resistance and reactance

E9G05 Which of the following is a common use for a Smith chart?
Determine the length and position of an impedance matching **stub**

E9G06 On the Smith chart shown in Figure E9-3, what is the name for the large outer circle on which the reactance arcs terminate?
Reactance axis

E9G07 On the Smith chart shown in Figure E9-3, what is the only straight line shown?
The resistance axis

E9G08 What is the process of normalization with regard to a Smith chart?
Reassigning impedance values with regard to the prime center

E9G09 What third family of circles is often added to a Smith chart during the process of solving problems?
Standing wave ratio circles

E9G10 What do the arcs on a Smith chart represent?
Points with constant reactance

E9G11 How are the wavelength scales on a Smith chart calibrated?
In fractions of transmission line electrical wavelength

SUMMARY – ANTENNAS AND TRANSMISSION LINES

E9H Receiving Antennas

E9H01 When constructing a Beverage antenna, which of the following factors should be included in the design to achieve good performance at the desired frequency?
It should be one or more wavelengths long

E9H02 Which is generally true for low band (160 meter and 80 meter) receiving antennas?
Atmospheric noise is so high that gain over a dipole is not important

E9H03 What is Receiving Directivity Factor (RDF)?
Forward gain compared to average gain over the entire hemisphere.

E9H04 What is an advantage of placing a grounded electrostatic shield around a small loop direction-finding antenna?
It eliminates unbalanced capacitive coupling to the surroundings, improving the nulls.

E9H05 What is the main drawback of a small wire-loop antenna for direction finding?
It has a bidirectional pattern

E9H06 What is the triangulation method of direction finding?
Antenna headings from several different receiving locations are used to locate the signal source

E9H07 Why is RF attenuation used when direction-finding?
To prevent receiver overload which reduces pattern nulls

E9H08 What is the function of a sense antenna?
It modifies the pattern of a DF antenna array to provide a null in one direction

SUMMARY – ANTENNAS AND TRANSMISSION LINES

E9H09 What is a Pennant antenna?
A small, vertically oriented receiving antenna consisting of a triangular loop terminated in approximately 900 ohms

E9H10 How can the output voltage of a multiple-turn receiving loop antenna be increased?
By increasing the number of turns and/or the area

E9H11 What feature of a cardioid pattern antenna makes it useful for direction finding?
A very sharp single null

E0 – SAFETY

[1 exam question – 1 group]

E0A Safety

E0A01 What is the primary function of an external earth connection or ground rod?
Lightning protection

E0A02 When evaluating RF exposure levels from your station at a neighbor's home, what must you do?
Ensure signals from your station are less than the uncontrolled Maximum Permitted Exposure (MPE) limits

E0A03 Over what range of frequencies are the FCC human body RF exposure limits most restrictive?
30 to 300 MHz

E0A04 When evaluating a site with multiple transmitters operating at the same time, the operators and licensees of which transmitters are responsible for mitigating over-exposure situations?
Each transmitter that produces 5 percent or more of its MPE limit in areas where the total MPE limit is exceeded.

E0A05 What is one of the potential hazards of operating in the amateur radio microwave bands?
The high gain antennas commonly used can result in high exposure levels

E0A06 Why are there separate electric (E) and magnetic (H) field MPE limits?
The body reacts to electromagnetic radiation from both the E and H fields
Ground reflections and scattering make the field strength vary with location
E field and H field radiation intensity peaks can occur at different locations

All these choices are correct

E0A07 How may dangerous levels of carbon monoxide from an emergency generator be detected?
Only with a carbon monoxide detector

E0A08 What does SAR measure?
The rate at which RF energy is absorbed by the body

E0A09 Which insulating material commonly used as a thermal conductor for some types of electronic devices is extremely toxic if broken or crushed and the particles are accidentally inhaled?
Beryllium Oxide

E0A10 What toxic material may be present in some electronic components such as high voltage capacitors and transformers?
Polychlorinated biphenyls

E0A11 Which of the following injuries can result from using high-power UHF or microwave transmitters?
Localized heating of the body from RF exposure in excess of the MPE limits

~~~End of question pool text~~~

NOTE: Graphics required for questions in sections E5, E6, E7, and E9 are on the following pages.

Figure E5-1

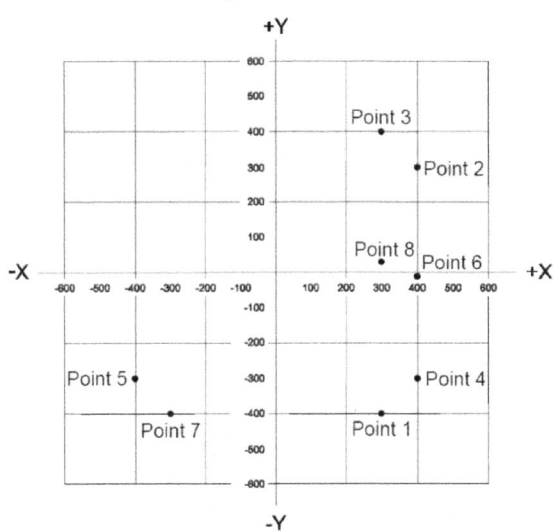

Figure E6-1

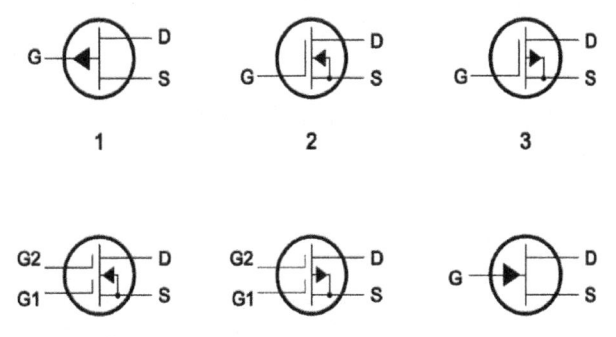

Figure E6-2

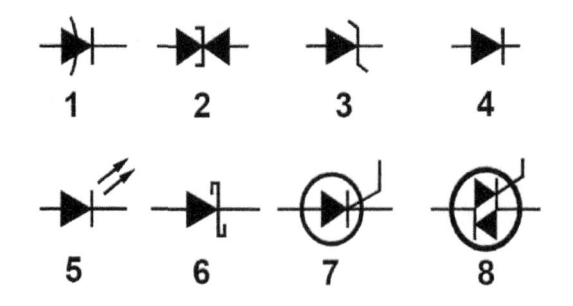

1 2 3 4
5 6 7 8

Figure E6-3

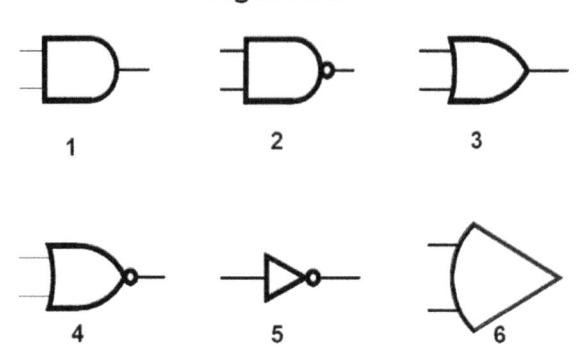

1 2 3
4 5 6

Figure E7-1

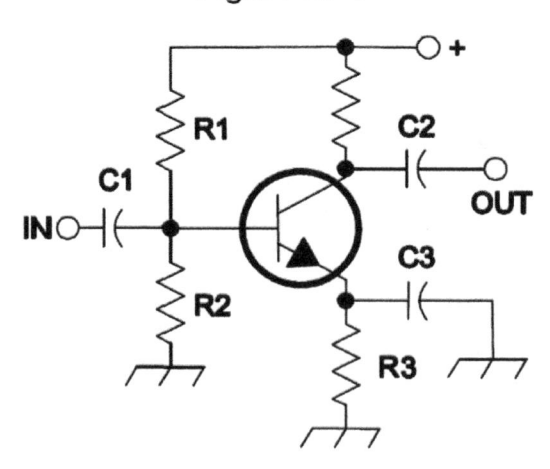

Page 190   Extra Class – The Easy Way

Figure E7-2

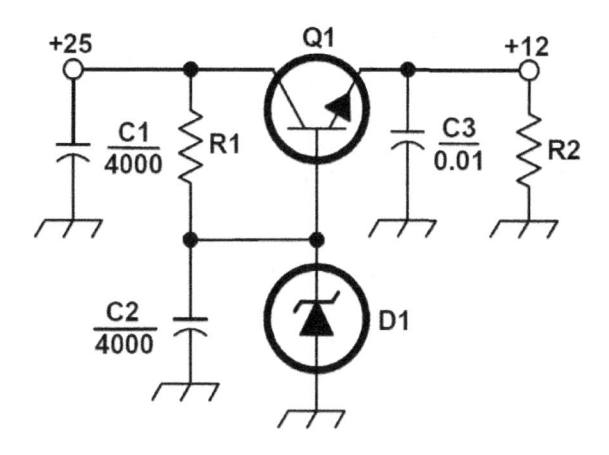

Figure E7-3

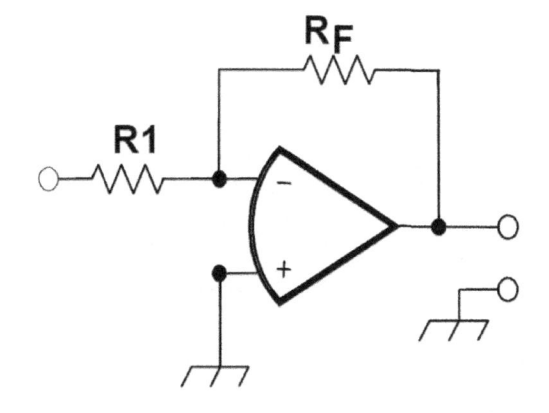

Extra Class – The Easy Way

Figure E9-1

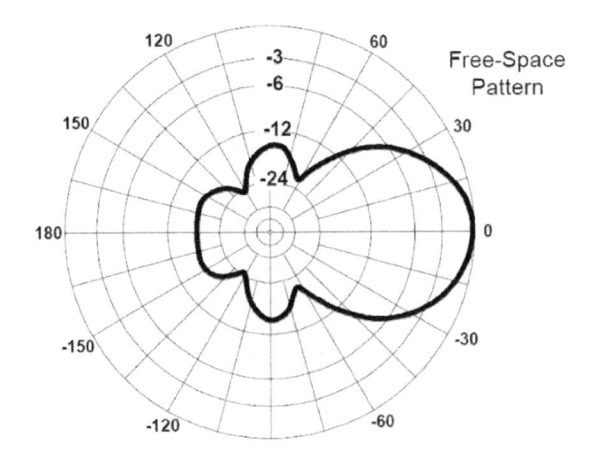

Free-Space Pattern

Figure E9-2

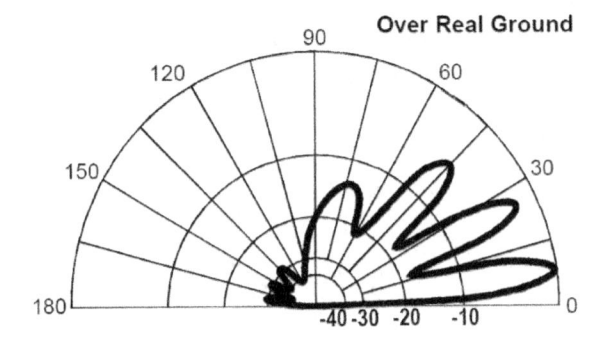

Over Real Ground

Figure E9-3

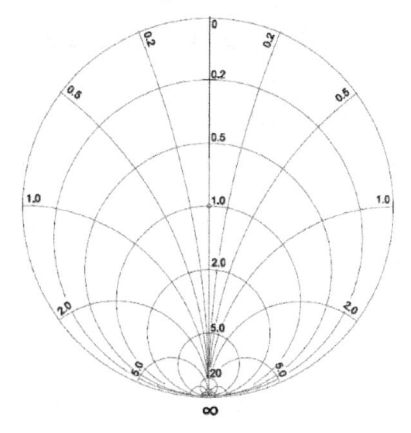

BONUS MATERIAL
LEARNING CW

CW is not required for any class of license, but learning it is not that difficult. Think of CW as a foreign language with 26 words and count to 10. Would you think mastering that an impossible task? Probably not. I learned Morse code, and I can't memorize anything. Thankfully, you aren't memorizing. You are training to recognize a sound.

Boy Scouts was my first exposure to CW. I had to relearn it as a Ham. What went wrong? In Boy Scouts, we memorized CW as dots and dashes. The decoding was visual, not audible. You had a little chart and looked up every letter. Your brain translates what it sees or hears into dots and dashes and then translates those into letters. That is a multi-step process. It is like learning to translate from English to Spanish to get to French.

Radio CW is audible. Learn to recognize the sound, not memorize dots and dashes. "A" is not short-long or dot-dash. It is not even dit-dah. It is the sound made by dit-dah.

For the same reason, do not learn that "A" sounds like "Ah-pull" or that the letter "A" has a short line and a long line. I've seen pictograms like a child's alphabet block with a picture of a bee for "B." These gimmicks introduce additional mental steps. Now you are going from English to Spanish to French to get to Russian. Learning the sound eliminates all the in-between translations.

Learn CW by hearing it one or two letters at a time until you can immediately make the connection from the sound to the letter. Then you go on to more letters and build. This is called the Koch method.

LEARNING CW

Another impediment to learning was the way we sent CW. At slow speeds, "A" became the sound diiiiiiit-daaaaaaaaaah. Then, the following letter came immediately with no time for the brain to decode. The modern way is called Farnsworth timing.

Farnsworth timing sends the letters at 16-20 words per minute, [5] so each letter has one distinct sound. Individual dits and dahs do not form a letter. You should not try to hear the different elements; the letter is one sound.

To slow down the pace, Farnsworth increases the space between letters and words. Increasing the space does two things. It reinforces a single sound as the letter, and it gives the brain extra time between letters to do the translation. You should learn to send using Farnsworth timing. That is how the other guy learned to receive. I set my keyer around 22-26 words per minute and slow down by increasing spaces.

After you recognize the letters, you will recognize words. When you read, you do not see letters; your brain knows the words. "Word" is not "W-O-R-D." The same happens with Morse code. You learn to recognize your call sign, RST, 5NN, TU, 73, and other common "words" without thinking about the individual letters or the elements that make up the letters.

Learning Code takes practice. In the beginning, listen to a Code practice CD or audio file. K7QO offers a free course download on his website, K7QO.net. G4FON has a Koch trainer at G4FON.net. There are other sources, as well.

---

[5] Words per minute (WPM) is based on a 5 letter word. "Paris" is an often-used standard. Send "Paris" 20 times in a minute for 20 WPM.

LEARNING CW

Once you get the letters down (mostly), listen to some QSOs, so you learn the standard QSO pattern. Then, GET ON THE AIR! There is no better practice than making actual contacts. Real QSOs are exciting and won't seem like tedious practice.

FISTS CW Club and Straight Key Century Club (SKCC) promote CW, and you will find slow CW on the frequencies suggested on their web pages. Both FISTS CW Club and SKCC assign you a member number you exchange with other members to collect awards. It is a fun challenge.

Suggested frequencies to find slower CW include:
3.550 – 3.570 MHz   21.055 – 21.060 MHz
7.055 – 7.060 MHz   28.055 – 28.060 MHz
14.055 – 14.060 MHz
You can also venture into the old Novice Bands and find slow CW.

At first, concentrate on the QSO Trinity: RST, QTH, name. You know the pattern, so you know what to expect. With time, you can venture into more complex conversations. Try to match the speed of the other guy, but if you can't, "QRS" means "slow down," and "QRQ" means "speed up."

For more challenging practice, tune in to the W1AW Code Practice Sessions. These are texts from *QST* magazine, and because the words are longer and not always as predictable, harder to copy. ARRL offers code proficiency certificates that will look beautiful on your wall. http://www.arrl.org/w1aw-operating-schedule will tell you when and where to listen.

INDEX

304A, 31, 128
Acceptor, 52, 146
Adaptive filter, 73, 163
AFSK, 24, 85
Alpha cutoff, 53
Analog-to-digital, 32, 36, 37, 73, 74, 80, 81, 130, 133, 163, 164, 168, 169
Antenna bandwidth, 88, 175
Apparent power, 49, 50, 144
APRS, 24, 122
ARQ, 84, 171
Ascending pass, 18, 117
ASCII, 83, 85, 86
Astable multivibrator, 63, 155
Aurora, 28
Automatic Link Enable, 26, 124
Balanced modulator, 72, 162
Band edge, 7, 106
Baseband, 72, 82, 162, 170
Baudot, 85, 86, 173
Beam-width, 89
Beryllium oxide, 105, 188
Beta, 52, 146

Beverage antenna, 102, 185
BiCMOS, 56, 149
Bipolar transistor, 52, 66, 146, 157
Blocking dynamic range, 38, 135
Brute-force AC-line filter, 40, 137
Bz, 30, 128
Cabrillo, 22, 120
Capture effect, 36, 133
Cardioid, 92, 103, 177, 186
Cavity filter, 68, 159
CEPT, 11, 110
Charge controller, 70, 160
Chebyshev filter, 67, 158
Chordal hop, 29, 127
Chroma, 20, 119
Circularly polarized, 19, 28, 118, 126
Class A, 64, 156
Class AB, 64, 156
Class C, 65, 156
Class D, 64, 156
Class X, 30, 128
CMOS, 56, 149
Colpitts, 77, 166
Common mode, 40, 137
Comparator, 55, 148

INDEX

Contact diode, 54, 148
Crystal lattice filter, 68, 159
Decade counter, 63, 155
Decimation, 74, 163
De-emphasis, 72, 162
Delta matching, 96, 180
Depletion-mode FET, 53, 147
Desensitization, 38, 39, 135, 136
Deviation ratio, 82, 170
Digital time division multiplexing, 83, 170
Diode envelope detector, 73, 162
DIP, 59, 61, 152
Direct digital conversion, 73, 74, 163, 164
Direct digital synthesizer, 78, 167
Direct sequence, 84, 172
Dither, 80, 168
Drop-out voltage, 71, 161
Effective radiated power, 8, 19, 87, 89, 107, 118, 174, 175
Elevation pattern, 90, 93, 178

Elliptical filter, 67, 158
Elliptically polarized, 29, 127
EME, 23, 27, 122, 125
Far field, 91, 93, 176
Faraday rotation, 19, 118
Fast Fourier Transform, 74, 163
Fast-scan, 20, 118, 119
Ferrite cores, 58, 150
Finite Impulse Response, 75, 164
Flip-flop, 63, 155
Foam dielectric, 99, 182
Folded dipole, 93, 177, 178
Forward Error Correction, 25, 83, 123, 170
Fourier analysis, 80, 168
Frequency discriminator, 72, 162
Front-to-back ratio, 90, 94, 176, 179
Front-to-side ratio, 90, 176
FSK, 24, 25, 26, 123, 124
G5, 31, 128
G5RV, 93, 178
Gallium arsenide, 52, 59, 146, 151

INDEX

Gamma match, 96, 180, 181
Gray code, 84, 171
Grounded-grid, 66, 158
Hairpin matching, 96, 181
Hartley, 77, 166
Hepburn maps, 27, 125
Hilbert-transform filter, 73, 163
Hysteresis, 55, 148
IMD, 35, 85
Infinite Impulse Response, 75, 164
Interlaced scanning, 20, 119
Intermodulation, 32, 35, 38, 39, 41, 66, 85, 130, 132, 135, 136, 157, 173
Isotropic, 87, 89, 91, 107, 174, 175, 176
JT65, 23, 24, 122
L band, 19, 118
Ladder line, 98, 182
LED, 54, 61
Letter B, 45, 141
Line A, 16, 115
Linear transponder, 18, 19, 117, 118
Loading coil, 94, 95, 179
Logic analyzer, 33, 131
Long path, 29

Low pass filter, 64, 81
Magnetizing current, 59, 150
MDS, 36, 133
Meteor scatter, 23, 27, 121, 126
Method of Moments, 91, 176
Microphonic, 78, 166
MMICS, 59, 60, 151, 152
Modulation index, 12, 81, 82, 111, 169
Monostable multivibrator, 63, 155
MOSFET, 53, 54
MPE, 104, 105, 187, 188
MSK144, 23, 121
NAND gate, 56, 63, 149, 155
National Radio Quiet Zone, 9, 108
Noise blanker, 40, 41, 137
Noise figure, 36, 60, 133, 152
Noise floor, 36, 38, 133, 135
NOR gate, 57, 63
Normalization, 101, 184
NOT operation, 57, 149
NP0, 78, 166
NTSC, 20, 118, 119
OCFD, 92, 177

INDEX

Op-amp, 75, 76, 77, 164, 165
Optical shaft encoder, 61, 153
Optoisolater, 61, 62, 153
OR gate, 63, 155
Orthogonal Frequency Division Multiplexing, 82, 170
Oscilloscope, 32, 33, 130, 131
PACTOR, 25, 123
Parabolic dish, 94, 178
P-channel junction FET, 53, 147
PEP, 80, 107, 168
Permeability, 58, 150
Phase angle, 46, 47, 50, 141, 142, 143, 144, 145
Phase noise, 35, 37, 133, 134
Phase-locked loop, 79, 167
Phasor, 47, 142
Photoconductive, 61, 153
Photovoltaic, 61, 62, 153, 154
Pierce, 77, 166
Piezoelectric effect, 58, 150

Pi-L network, 67, 68, 158
PIN diode, 54, 148
Polar coordinates, 46, 47, 142, 143
Polychlorinated Biphenyls, 105, 188
Powdered-iron, 58, 150
Power factor, 49, 50, 144, 145
Pre-emphasis, 72, 162
Prescaler, 32, 130
Pre-selector, 35, 133
Product detector, 73, 162
PSK31, 26, 83, 124, 171
Pull-up, 56, 149
Push-pull, 64, 65, 156
Q, 34, 42, 43, 44, 68, 76, 79, 94, 95, 132, 139, 140, 159, 167, 179
Quadrature, 73, 163
RACES, 10, 109
Radio horizon, 30, 128
Ray tracing, 30, 127
Reactance modulator, 71, 72, 161
Reactive power, 50, 51, 144, 145
Receiving loop, 103, 186
Rectangular coordinates, 47, 142, 143

Page 200 Extra Class – The Easy Way

INDEX

Rectangular notation, 47, 142
Reflection coefficient, 34, 97, 132, 181
Resonance, 42, 43, 59, 131, 132, 139, 151
RF ground, 95, 180
Rhombic, 93, 177
S band, 19, 118
S parameters, 34, 132
S11, 34, 132
S21, 34, 131
SAR, 105, 188
Satellite's mode, 18, 117
Saturation, 57, 64, 66, 150, 151, 156, 157
Sawtooth wave, 80, 168
Schottky diode, 54, 147, 148
SDR, 36, 37, 74, 133, 134, 164
Self-spotting, 21, 120
Series regulator, 69, 160
Shunt regulator, 69, 160
Silicon, 52, 62, 146, 147, 154
Sinusoidal, 83, 171
Skin effect, 49, 95, 143
Smith Chart, 100, 101, 183, 184
Soil conductivity, 88, 175

Solar flare, 30, 31, 128, 129
Solid state relay, 61, 153
Special Temporary Authority, 16, 115
Spectrum analyzer, 32, 35, 130, 132
Sporadic E, 29, 127
Spread spectrum, 16, 17, 84, 85, 115, 116, 172
Spurious emission, 9, 12, 17, 108, 111, 116
SSTV, 19, 20, 21, 108, 117, 119, 120
Step-start, 71, 161
Stub match, 96, 180
Successive approximation, 80, 168
Surface mount, 60, 152
Susceptance, 45, 141
Symbol rate, 83, 84, 170, 171
Takeoff angle, 93, 94, 178
Taps, 75, 164
Telecommand, 12, 13, 14, 112, 113
Telemetry, 12, 13, 111, 112
Thermal runaway, 66, 157
Third-order intercept, 39, 136

INDEX

Time constant, 44, 141
T-network, 67, 158
Transequatorial propagation, 28, 29, 126
Triangulation, 102, 185
Tri-state logic, 55, 148
Tropospheric, 27, 28, 125, 126
Truth table, 64, 156
Varactor diode, 54, 147
VEC, 14, 15, 113, 114
Vector network analyzer, 34, 132
Velocity factor, 97, 98, 99, 181, 182
Vertical Interval Signaling, 21, 119
Vestigial sideband, 20, 119
Wilkinson divider, 97, 181
Wire-loop, 102, 185
Zener diode, 54, 69, 147, 160
Zepp, 93, 178